THE
TIDE

In Search of Sir Thomas Browne

Anatomies

Periodic Tales

British Design

Panicology (with Simon Briscoe)

Findings

Zoomorphic

The Most Beautiful Molecule

World Design

New American Design

THE
TIDE

THE SCIENCE AND STORIES
BEHIND THE
GREATEST FORCE ON EARTH

HUGH ALDERSEY-WILLIAMS

W. W. NORTON & COMPANY

Independent Publishers Since 1923

NEW YORK • LONDON

First published in Great Britain under the title *Tide: The Science and Lore of the Greatest Force on Earth*

Manufacturing by Quad Graphics Fairfield
Book design by Mary Sarah Quinn
Production manager: Anna Oler

ISBN 978-393-24163-1

W. W. Norton & Company, Inc.
500 Fifth Avenue, New York, N.Y. 10110
www.wwnorton.com

W. W. Norton & Company Ltd.
Castle House, 75/76 Wells Street, London W1T 3QT

1 2 3 4 5 6 7 8 9 0

TO JOHN

CONTENTS

LIST OF ILLUSTRATIONS

AUTHOR'S NOTE

In general, I use metric (SI) units, especially when discussing scientific measurements. But I also use familiar units where to do otherwise would seem perverse. Some units—the foot, the mile—have a rightness about them that transcends international systems of measurement.

I KNOT = I NAUTICAL MILE PER HOUR

= 1.15 STATUTE MILES PER HOUR

= 1.85 KILOMETERS PER HOUR

= 0.51 METER PER SECOND.

Historical tide data given at the head of various chapter sections are calculated using the United Kingdom Hydrographic Office Admiralty EasyTide website, www.ukho.gov.uk/easytide, except for the datums for Dover, which are from the source indicated in that chapter, and the one for Stockholm, which is from Martin Ekman.

Dates are generally given according to contemporary records, but tidal calculations for Old Style (Julian calendar) dates have been adjusted.

THE
TIDE

INTRODUCTION

THIS IS NOT A BOOK ABOUT THE SEA. It does not feature long days before the mast, scurvy, whale boats, pirates, ship's biscuits, and tots of rum. It does not brave hurricanes or typhoons. It does not center on a man condemned forever to sail the oceans, nor does it have mermaids to lure him astray. It is not about the vasty deep, man's battle with it, and his often curious reasons for going into that battle in the first place.

This is a book about the sea. It is about the sea that we all know—the beach, the coast, land's edge, land's end. The sea we regard as our vacation playground, yet understand hardly at all. The sea that we move cautiously upon, and that moves us in mysterious ways, both physically and emotionally.

The tides are complicated, and some people find they obtain their most elegant explanation in the form of mathematics. Yet even if you are able to interpret the symbols and equations, you immediately lose the visceral sense of the oceans' rising and falling, the sense of the tide's upper hand over humankind's maritime adventuring. The tide has simple physical power over small craft at sea, but it also has power over the senses and the mind. To stand and gaze out from aboard a boat held at anchor upon a tidal body of water within sight of land is to subject oneself to a hallucination of bewildering intensity. For, during those regular periods when the tide is running strongly in one direction or another, it will seem that the boat is cleaving purposefully through the sea, even though reason tells you it is going nowhere. It is a powerful and disturbing illusion, enhanced in its hypnotic effect perhaps by sunbeams bouncing off the streaming water or the rhythmic undulation of the boat, or by your own hunger or thirst. You blink hard to dismiss it, but when you open your eyes it is still there—the boat is definitely moving forward, anchor chain and all, this last determinedly probing the waters ahead like a narwhal's tusk.

After a while, your brain reframes the illusion. Mesmerized by the scintillating wavelets, it now conceives that it is the water that is static while you and the boat are racing together with the distant land toward some shared destination. The surface of the water may be glassy and smooth, wrinkled only slightly by the deep turbulence of the tidal flow itself, or it may be choppy, as it is when the wind blows against the tide and friction between the air and the water kicks up short, steep waves. It doesn't matter. Your impression nevertheless is that all this water must be essentially stationary. For what great power could convey such a mass of ocean back and forth so swiftly without apparent effort or cause?

We learn at school that the main answer to this question is the moon and its gravitational pull on the earth, and by adulthood we have assimilated this information without demur. We stupidly take the force of gravity as read, and reserve our wonder for more modern oddities, such as quantum theory or dark matter. Yet how very odd the colossal, invisible force of gravity still is if we stop to think about it at all. With the rushing tide, we have a visible, soaking, undeniable expression of that weirdness.

Before we understood them in a scientific sense, the tides were already comprehended in their way by mythmakers and storytellers. The actual power of the tide to drag sailors to their death is surely enough to explain the lure of the siren and the sucking tentacles of the kraken. The sea has no need of assistance from malevolent creatures, which are the mere invention of ignorant seafarers, made up because they offer a more believable story than the horrid idea that something as routine and automatic as the unaided tide has the power to take a person's life. The fabulous sea monsters that adorn the peripheries of old explorers' maps may similarly be unfamiliar deep-sea species whirled up into the light by tidal upwellings, while the tidal bores of various rivers earn themselves the names of gods and monsters.

Scientific knowledge offers an alternative explanation for some of these stories, but the stories live on. Science has begun to investigate the mysteries of the oceans—only just begun, really: oceanography is one of the youngest of the sciences, and the subject is all too obviously a large one. The investigation of the oceans has started, but we can hardly be said to have tamed them. We can now predict the tide theoretically to a high degree of accuracy—far more accurately, in fact, than it ever occurs in our real experience, where other factors interfere. Why the drive to know in such obsessive detail? In part, it is because the tides provide a perfect example of the complex problem that should be exactly soluble. All the variables are known; it is only a matter of doing the sums. The tide offers an irresistible mathematical tease, which is undoubtedly why it has historically attracted some of the world's finest physicists and astronomers. But there are also practical reasons why it is important to have precise answers beyond the needs of navigators and fishermen, and these turn out to have relevance for the future of us all.

There is a good reason why, despite all this, no accessible book about the science of the tides has been published. The reason is that the topic swiftly becomes more complex than lends itself to explanation in words. The author of a field guide to the salt marshes of New England gave his drily brief synopsis of the subject this heading: "Tides in Perhaps More Detail Than I Should Include in This Book." I know how he feels. Harmonic equations give the scientist a far more versatile tool for understanding what goes on. But I know that I cannot follow this route—for your sake, and for mine.

Instead, I have tried a different approach. I give an episodic history of the science of tides from the earliest times up until the present day. This course allows me to voyage from the earliest science of Aristotle, who is said to have drowned himself when he failed to figure out the Greek tides, to the better-informed investigations of Galileo and Newton, and then on to a scientific understanding so complete that we are now able to predict the tides with that high degree of theoretical accuracy (far greater than any sailor needs), which is yielding important new evidence of our own impact on this watery planet. Along the way, we drop anchor in less familiar harbors too, pausing to acknowledge the unexpected contribution to the understanding of the tides by figures such as Bede and Saint Thomas Aquinas, men who are not primarily thought of as scientists, but whose great minds could not ignore this cosmic puzzle.

I weave the scientific strand of my narrative with two others: one in which stories of events—be they historical, artistic, or entirely fabulous—where the tide plays a crucial role get their due; and another, in which I go in search of special places made by the tide. I include these episodes to show that the tide is not only a scientific challenge, but also a force of a quite different kind—a physical and a psychological influence on our culture whose presence cannot be denied. The tides have determined the course of battles and have inspired poets and artists. And they continue to do so today.

In weaving these strands, strict chronology is occasionally sacrificed for the sake of a thematic connection. The science is, I hope, made simple but not simplistic. This is not a textbook about the tides. It is a book of stories and journeys. I hope you find yourself able to go with the flow. It is always unwise to fight the tide.

WHICH CAME FIRST—TIME OR TIDE? It is impossible to arrive at a clear answer. Both words have Anglo-Saxon roots. Phonetically, our "tide" is linked to the German *Zeit*, which still means "time," while the newer German word *Gezeiten* now denotes the "tides of the sea." In early English, "tide" was not especially a quality of the sea, but more a way of describing a significant time—a usage we retain in the archaic suffixes of events in the church calendar such as Shrove*tide* and Whitsun*tide*. The Old English *heahtid*, or "high tide," had the simple terrestrial meaning of a festival or high day. However, for those whose livelihoods depended on the sea, the most important "tide" was always the day's high or low water.

Some sources indicate that "tide" only came to refer chiefly to the rise and fall of the sea in the fourteenth century, when this medieval neologism took its place alongside "ebb" and "flood," or *ebba* and *flod*, which stem from Old Norse and proto-Germanic, and which have still older and more distant roots in Indo-European languages. The flood is simply the incoming tide, and it is clear there is no intrinsic conflict between this meaning, exclusively to do with the sea, and the more general meaning of the word as a threateningly raised level of water from any source. The ebb is the receding tide. The figurative senses both of "flood," as in a flood of tears, and of "ebb," as a kind of last chance, still in widespread rhetorical usage today, were also established by the fifteenth century.

In his tidal adventure *Passage to Juneau*, Jonathan Raban suggests that it was the invention of the compass that gave man the temerity to think he might sail in straight lines across the sea. The evidence of this boldness is etched in medieval portolan charts showing sailing directions, crisscrossed with dozens of dead-straight rhumb lines. By contrast, a more natural navigator in tune with the elements would simply feel the currents under his hull and use the tides to advantage without worrying what his course might look like on some hypothetical map. In like fashion, it was perhaps only the invention of the clock that finally enabled the more abstract concept of *time* to supplant the earthly *tide* as many people's measure of days.

We are left with only a few linguistic relics of the age before technology forced time and tide apart. If, for instance, in colloquial usage, somebody offers to "tide me over," it means they will support me for a time, usually with money. But the original meaning of the phrase is nautical. To tide a boat over a sandbar, for example, is to use the period of high water to get past shallows that would be impassable at other states of the tide.

With their easy assonance and their entwined historical usage, it is no great surprise to find "time" and "tide" forced together in memorable sayings. The aphorism "time and tide wait for no man" is the best known of these. It sounds like Shakespeare, or perhaps Chaucer. But it is even older, attributed by some to a shadowy Saint Marher from the early thirteenth century. It contains both a truth and a deception. For, of course, it is true that both time and tide are governed by celestial laws and so lie beyond the control of man. But whereas time passes forever, the tide always returns. Time is a continuum, the backdrop against which all things happen; the tide is always an event in time.

The senses in which time and tide do not wait are thus subtly different. If we miss an appointment because time does not wait, the moment has passed and may never come again. If we miss a tide, there is the same sense of an opportunity not taken, but there is also implicit in this expression a fairly certain knowledge that the opportunity, or something very much like it, will re-present itself in due course.

And there are other important differences. Time's arrow is weightless, but the tide has massive force behind it. Its flux and reflux are sensible. As we stand in the waves, the flood smacks our chest, the ebb drags on our shins. Struck by these sensations, we momentarily forget that the tide is a cyclical phenomenon, in which each motion is endlessly repeated with infinite small variations, and feel just the singular event. This visceral power is reflected in our adaption of the tide's special vocabulary—the word "tide" itself, as well as "ebb" and "flow"—to broader linguistic purposes.

Thus the high tide may bring unique opportunity, as Shakespeare reminds us in the famous speech of Brutus on the eve of the Battle of Philippi:

There is a tide in the affairs of men,
Which, taken at the flood, leads on to fortune;
Omitted, all the voyage of their life
Is bound in shallows and in miseries.
On such a full sea are we now afloat;
And we must take the current when it serves,
Or lose our ventures.

When the waters rise high enough, they enable the voyage of (self) discovery to begin.

THERE IS, MEANWHILE, a desolation about the ebb, a physical emptiness that is echoed by a psychological emptiness, the loss of seawater equated in our primitive memory with loss of the water of life. The water rushes out to join the ocean; it runs away to sea, in fact. The action has about it some of the naïveté and desperation implicit in that expression. It is bound on its destiny and leaves behind it a bereft expanse of sand or mud.

With the ebb goes life and life's chances, or so it might seem to us. But nature doesn't regard it like this. Nature's richest seams are often where two habitats chafe together: birds often prefer hedgerows, not open fields or the heart of the forest; amphibians need the water's edge, not deep water or parched earth. Margins provide choice—food to one side, perhaps, refuge to the other. And where that margin is always shifting about, as it is on the tide's edge, within that glistening ribbon of land that lies between the high-water mark and the low, the stock is constantly replenished with new riches. This intertidal zone, which stretches and shimmies its way for perhaps a million kilometers (it depends how you measure it) in continuous lines around all the continents, contains some of the world's most biodiverse regions. And what's more—according to the lyrics of Rodgers and Hammerstein's *Pipe Dream*, perhaps the only musical ever to feature a marine biologist in the central role, adapted from John Steinbeck's stories of Cannery Row—it crawls with some of the very weirdest of creatures.

It is the regular rise and fall of the sea that restocks the shelves of the world's continents twice a day more efficiently than any supermarket. Nature is organized on the principle of conserving energy. The objective of all organisms is to survive and propagate while making the minimum effort to do so. If a migration can be accomplished by hitching a ride, if food can be delivered to your doorstep for free, then so much the better. This is the essential service that the tide provides—a vast source of physical energy ready to be tapped by any animal or plant that can adapt to its demanding schedule.

This is a book about the discovery and science of the cosmic rhythm that governs our planet. But it is also a book about places. Along that nearly infinite shoreline of the earth's continents lie points where the irresistible force of the tide and the immovable matter of the earth converge in such a way as to create special sites—natural harbors, river mouths, firths and fjords, isthmuses and promontories—that take the particular shape they do because of the sculpting action of the tide.

For the seagoing few, the tide concocts other, more terrifying places—whirlpools, tidal races, and overfalls of destructive waves, as well as locations of huge tidal range and others, far out in the ocean, where, even though tides swirl all around, the sea level never varies. These places may not appear in landlubbers' atlases, but they are indicated on nautical charts—a few of them at least. They are places, but they are also theatrical events. They occur only in these special places at scheduled times, and then only if the actors feel like turning up. A spectacular tidal whirlpool or river bore may arise on one tide but not on another one equally high, simply because local and temporary factors conspire to prevent it.

These places of excess and anomaly arise wherever the rise and fall of the ocean is suitably constrained by the local shape of the seabed and coastline. They may be equatorial or polar, stormy or placid, populous or remote. I chose to travel to Nova Scotia in Canada, where the tides are the greatest in the world, although I might have visited the coasts of Argentina or northwestern Australia, where they are nearly as great. I saw surging currents and vast whirlpools in Arctic Norway, although I might have seen them in Japan or the Strait of Magellan or Vancouver Bay. It matters not. These features are replicated in some degree on every coast of every ocean.

Furthermore, I can accomplish much of my task closer to home. For it happens that the coast of the British Isles is one of the most tidally lubricated coasts anywhere in the world. Surely no coastal nation has, when the velocities and vertical ranges are measured along its entire length, more tide than Britain. What does this mean? More washes up, more washes away; there is more damage, more erosion, more sediment, more life, more death. We kid ourselves that we are a maritime nation, yet our ignorance of all this activity going on ceaselessly around our shores is quite astonishing.

The tide's evanescent places don't leave scars, create tracks, or memorialize themselves, as famous roads or mountain peaks do on land. They live through seeing and telling only. This is why my history of the science of the tide is interwoven with tales of the fantastic and with my own travels.

Finally, if a state of the tide has the power to make a place, then a whole tidal cycle must trace a path. Simply by remaining static in one coastal spot and watching the tide, I should find myself transported into new terrain and then back to where I began with such magical completeness that I would be left to wonder whether what I had seen was truth or illusion.

I decided this was how I would begin. I have seen coastal places at high water and at low. I have used the sea at both extremes of the tide—around high water for sailing in the shallows, and at low water for rock pooling. But now I realize I have never taken due note of the whole natural cycle. I have never seen *exactly* what happens when the tide moves. Perhaps I could understand the tide intuitively by communing with it, as well as, in a theoretical way, by grappling with its science. At the least, I thought, the experience would reveal the questions I should ask.

Let us take our seats. Nature's greatest marine drama is about to begin. The performance will last approximately twelve hours and thirty minutes.

1

TO VIEW
THE LAZY TIDE

THIRTEEN HOURS

BLAKENEY, NORFOLK

	HW	LW	HW	LW
September 4, 2013	05:57		18:27	
	3.0 m		2.9 m	

It is an odd idea, I admit, simply to sit and watch the water for twelve or thirteen unbroken hours. You might find similes coming to mind to do with watching paint dry or grass grow. But I will shut these unhelpful analogies out of my mind. I do not know what I might see, but I will at least try to note down anything I do. I do not know what I might see, and that will be the best of it.

The first requirement was to select a site where I could do this. Every part of the British coast is subject to substantial tidal movement. I live in Norfolk, a county that bulges obscenely out into the North Sea (in old satirical cartoons that depict Britain as a person, Norfolk is always the rump). The coast is correspondingly distended, and so I was spoiled for choice. I considered Blakeney Quay. I'd seen the tide running in there so fast around the bend in the river—I reckoned its speed as about 3 meters per second, based on counting as pieces of seaweed hurried by—that it sent thick wooden mooring posts into frenzied vibrations. But the place was too overrun with tourists, and I could see that I would be constantly interrupted by curious busy-bodies. Instead, I selected a site a mile or two away where I knew I would be undisturbed.

I could have selected many other favorite places for my task: the tidal River Yar that all but severs the Isle of Wight in two, where I spent my childhood summers; Griswold Point, at the mouth of the Connecticut River, introduced to me by my American cousin, where the rare piping plover scrapes its meager nest, or any of the deserted beaches I saw when I drove south through Oregon on Route 101 years ago; Cádiz, the peninsular Atlantic port of the Phoenicians and everybody since, where the city cats gather on the breakwater rocks, or nearby Cape Trafalgar, the sandblasted beach where I once stood contemplating the historic victory of the Norfolk hero Horatio Nelson over the Spanish fleet.

His own county was my choice in the end. The coastal landscape of Norfolk is predominantly flat. Low-lying farmland cedes to a broad margin of salt marsh infiltrated by a mazy network of muddy creeks before shingle banks and dunes raise a beach to the sea. The scene would be nothing like the domesticated sublime of the beach at Lyme Regis that Jane Austen describes in *Persuasion*, "where fragments of low rock among the sands make it the happiest spot for watching the tide." My prospect would be more like that in George Crabbe's epic poem of East Anglian life, *The Borough*. I would "view the lazy tide / In its hot slimy channel slowly glide." I would make myself into what Charles Dickens in *Our Mutual Friend* called one of "those amphibious human-creatures who appear to have some mysterious power of extracting a subsistence out of tidal water by looking at it."

Reading passages such as these, I saw that writers use the tide as a kind of hypnotist's watch. It is something to induce a state of reverie or, more dangerously, a trance. I would have to be careful not to fall into daydreaming if I was going to make more incisive observations of the unceasing rise and fall of the seas.

Next, I had to choose a suitable time of year and time of day to make my study. The tides are in constant action washing the world's shores, but they vary according to astronomical factors that are subject in turn to their own complex temporal rhythms. I did not want to freeze or fry out on the marshes, but more important than that, I would need my thirteen hours to fall during daylight in order to make my observations. Wherever you are, a full tidal cycle, from high water back to high water (or low to low), takes nearly this length of time. This constraint limited me to the months from March to September when the days were long enough. I also wanted to observe a fairly typical tide, not a huge one that would flush me out of my vantage point when high water approached, nor one so meager that I would miss the things I should normally expect to see.

Any thirteen-hour time slot guaranteed that I would see one high water, one low water, one full flood tide and one ebb. But where in the cycle did I want to start my work? This was more a matter of aesthetic preference and narrative design. To begin with the tide in full spate, either flooding or ebbing, seemed to me melodramatic. An obscure logic told me that low water would be a natural beginning: a bath or a bucket starts empty, after all, and its story is to be filled. This version would give the greatest sense of a flooding. I could watch the flood tide fill the creeks, but I would then have to see them empty again as the cycle came around, and something about this displeased me. Or, I could start at high water. But this was not right either: even though I would then end on a high, it seemed wrong to begin by witnessing the departure of the substance of my tale. I feared that the immediate ebb might be the end of my own story.

In the end, my choice was even more restricted. The tide table showed few days when the tidal range would be sufficient for my needs, the day long enough, the weather likely to be bearable, and the place quiet enough—a weekday during school term rather than a weekend—that I would not be disturbed. In the end, I chose a day when the sun would be rising just as the ebb was gathering pace. I would begin my observations about an hour after high water. The mood should be one of calm and expectation. My morning would see the tide recede and the muddy shore revealed. High water would come late in the day, and provide a well-timed climax. By starting an hour or two after high water, I would then stay on through the subsequent high water long enough to see the ebb begin again. This, I felt, would show more truthfully that the tidal cycle does not in any way peak or culminate at high water, as we might be tempted to think, but that it goes on in an eternal cycle in which no momentary state has any more claim to special status than another.

I would need to plan carefully to make the most of the day. I wanted to observe as much as I could that was related to the tide—the physical changes of the water, the dependent vegetation, the animals that came and went, the people who used it. I made a list of things to take: camera, notebook, graph paper, magnifying glass, bags for plant specimens. An old fiberglass windsurfer mast I transformed into a tide gauge by wrapping it at 10-centimeter intervals with bands of waterproof tape. I even loaded up my canoe onto the car roof in case I needed to make an aquatic foray for the further investigation of some mystery that flowed past.

Though I intended to be diligent in my observations, I imagined there might be long stretches when little was happening, and so I armed myself also with a copy of *The Oxford Book of the Sea*. It holds excerpts of many works I would need to familiarize myself with, from Rachel Carson's *Sea Around Us* to Matthew Arnold's allegorical poem "Dover Beach." These poems and prose pieces would remind me of the main to which my insignificant creek, thanks to the tide, was eternally connected and intermingled.

FIRST EBBING

I arrive on the marsh one warm September morning shortly after dawn. The weather is fine, and the sun has already burned off a land mist. A light southerly wind is beginning to ruffle the water. In the creeks that branch off the main channel, I can see from the movement of the slight film on the water's surface that the tide is beginning to run out.

I station myself by a wooden footbridge over one of the creeks, and set up my tide pole, pushing the foot of the mast under the water until I feel it meet the bottom, and then lashing the upper portion to the bridge handrail. I take my first reading: at 07:15 the water is 2.02 meters deep. Then I identify a suitable position from which to take photographs. The camera viewfinder includes the creek, which empties away from where I am standing into the main channel, as well as its banks, the bridge, and my tide gauge. In the foreground is a small launch that has lain on its mooring so long that it has grown a covering of lichen.

As the minutes pass, I notice that the ebb tide is already gathering pace. The wind is blowing into the creek, pushing ripples along it that disguise the smoother outflow of the tide. I had vaguely planned on taking readings every hour, but now I see that if I do this, I will be missing the story. Right now, I need to take readings every few minutes. At 07:30, the tide has already fallen to 1.92 meters.

I run back to where I have parked the car upstream, and unload the canoe so that I can bring it to my observation point while there is still enough water in the channel. I slip downstream hardly having to paddle. Looking at the water, the tidal flow is imperceptible even when I am sitting at its level; only the speed at which the mud banks and the moored boats rush past me gives me the clue. I am familiar with the hallucinatory feeling that this relative motion produces, but it still unnerves me every time. It is the sign of how uncanny we find it that the whole watery mass of the earth should be able to slide about according to its own mysterious rules on top of the solid base.

I tie up the canoe and return to my observation point. With the accelerating ebb, the encrusted boat in the creek has swung on its mooring, and its bow is pointing directly upstream—the clearest indication that the water is running out faster now. A wave has formed around the deepest bridge support as the water forges past. Its turbulence generates a succession of small whirlpools. They last for a few seconds each, commas and semicolons swept away by the current. I notice that they rotate both clockwise and counterclockwise. Why are they apparently immune to the famous bathroom myth that the water always swirls one way or the other as it disappears down the plughole? The larger whirlpools suck down small pieces of plant matter, churn them around for a bit, and then release them to the surface a few meters downstream.

Small islands of scum have materialized near the edges of the water. Where have they come from? Have they been lifted off the underwater mud by the flowing water? Have they been frothed up by the agitation of the tide? Are they animal, vegetable, or mineral? I realize I have no idea. They move off aimlessly like *îles flottantes*. Some of the *îles* saunter down the creek with the current, but others curl idly aside to explore tiny side channels, like pedestrians drawn by shop windows. I look closely at the water that is carrying them and see that they have been caught in eddies—momentary back-currents where a portion of water has been diverted from the main flow by some irregularity in the channel. They are a reminder that although the principles may be clear, and the tides at large subject to ideal prediction, there is in fact very little that is simple about the flow of the tides in real waterways.

Across the salt marsh, birds are beginning to feed. An argument of geese has blown up somewhere in the distance. Closer in, I hear the song of meadow pipits and the hollow peeping of waders. Two curlews bank steeply and land in the tributary nearest me, drawn, I suppose, by the promise of food in the freshly exposed mud. A swallow flies overhead, chattering as it goes. The swallows are beginning to gather on the wires for their migration—another natural cycle of coming and going governed by celestial clockwork.

It is already time for my next tide reading—my first official hourly one: 1.38 meters. I am amazed how much water has already drained out of the creeks, so quietly, without fuss or drama. There will be six hours of ebb, or there should be, and yet a third of it seems to have happened in just one hour. I do not know why this is.

I have seen a couple of early-morning dog walkers, and two or three boats have motored past, but now that the tide is well down and won't be back in for many hours, I am anticipating a quiet day. Once, this coast would have been busier at low tide, with local people gathering shellfish and samphire. But today the coast is a place of recreation, which depends on the water being high for its appeal. Few people come even for the scenery when the tide is out.

I decide to make a closer inspection of one of the *îles flottantes*. It is gray brown in color and frothy on top, like a disappointing cappuccino. The bubbles vary from pinhead to pea in size. Tiny insects ride on top—slim, long-legged flies with angular joints and elegant, backward-extending wings. While studying one of these, I notice further movement and see that the undersides of the *îles* are attracting their own insects, a mirror world of life under the water. I pick up some of the scum on my fingers, and it dissolves away. It is completely inoffensive, without discernible smell or taste. Clearly, it is a product of the natural world, and not the pollution we might guiltily be inclined to think.

Also attracted to the scum are some tiny segmented creatures, charcoal blue in color. They lie curled up, apparently inanimate. At first, I think they are seedpods. But when I pick one up, it springs to life, bustling around my finger. Its body is very supple, its rear end as well as its head craning and exploring its sudden displacement. It is so blue that I lose it when I place it on a bright plastic fender. When

I look it up later in my "complete" guide to British coastal wildlife, it is not there. The book, in fact, includes no life-form smaller than about 5 millimeters, and I wonder whether the publishers are catering to popular demand by including only organisms that are easily visible to the naked eye, and whether this prejudice is also reflected in the research priorities of marine biology.

I listen through the bird calls and the noise of distant traffic to a sound that I realize I have been dismissing until now as "the lapping of the water" or some such. But there are no waves to lap. So what is this sound? I stare intently at the mud that was exposed less than an hour ago and find that it is seething with small animals. Almost transparent worms have come to the surface and are gyrating in the film of water that adheres to the top of the mud. From time to time, the little holes from which they have emerged collapse or reopen with a wet pop, and it is the cumulative effect of these hundreds of openings and closings that makes the musical background murmur. The warming mud that has not seen the sun for half a day raises a sweet shellfish odor. The seabed is coming to life.

THE MUD AWAKES

The sun climbs higher. Bright dapples reflect from the rippling water onto the underside of the wooden bridge. Once, I recall, I was told there is a word for this delightful optical effect in the Venetian dialect, but I have never been able to track it down, and I wonder now whether it was the product of a dream. Perhaps I will discover the truth of the matter when I go to Venice to learn about the measures the city is taking to protect itself against tidal flooding.

I had thought that there would be longueurs in my day. But it is clear now that I will be kept very busy. I find it necessary to carefully plan my activity between each hourly tide reading, because I know I'll have the chance to do only certain things—like delving in the mud for worms, or observing how the wind whips up waves—at certain states of the tide. Suddenly, my schedule starts to look like a school timetable. I have the whole curriculum covered: plotting water level graphs (mathematics); observing mud life and marsh plants (biology); recording water flow (physics); canoeing (PE); contemplating the cosmic order of things (religious studies?). I will be so busy for the day that English will have to be canceled; *The Oxford Book of the Sea* lies unregarded, its pages turning crisp in the dry breeze.

By ten o'clock, my creek has virtually emptied. Opposite where I have chosen to sit is another little creek, which now reveals snugly fitted within it the rotting beams of some large wooden vessel long ago hauled in there to die. My tide gauge records 0.21 meter. Even in the main channel the ebb is now almost slack. The sign of this is that the mud stirred up by the turbulent earlier outflow, which still hangs in suspension in the water, is drifting downstream like smoke in the air.

Gulls have come down to feed on the glistening muddy bank. I watch one, a black-headed, and see something I have never seen before. The bird furiously paddles the surface of the mud with its feet. Then it stops and peers down. Then it paddles again. Eventually, I understand what it is doing. It is softening the mud with its red flippers in an effort to drum up some lunch.

At the mouth of my creek there is no longer anything to mark the tidal flow. A few little scum patches are held against the tide by the wind, making it impossible to guess at the water's movement beneath them. In the main channel, though, the tide is still falling visibly, and now audibly too. It laps past the tied-up boats. The water here is so murky that I cannot tell how deep it is, and I cannot easily investigate one thing I wished to: how the tide flows at different depths. If I were to track a certain cubic centimeter of water from the top surface of the high tide, would it go down until it reached the bottom, or would it go out sideways? Down or out? We talk about the tide being in or out, but the level of the water goes up and down. Which is it? Or is it a combination of the two? And which water flows the fastest? The water on the surface, where it meets least friction from the seabed? Or the momentous water in the bulk of the flow, at mid depth? Does my imaginary volume of water retain its cubic dimensions for long, or is it stretched out as it goes like a piece of bubble gum pulled from the teeth? Or, does it spread more complicatedly, tangling hydra-like with similar briny cubic centimeters on all sides? Naïve questions, perhaps. But I feel I should know. The idea that different parts of the same body of water can flow at different rates and in different directions seems counterintuitive at first, and yet it is clearly the case, because I know that while the tide is ebbing in my Norfolk creek, it is running in the opposite direction in similar creeks elsewhere, not even very far from here. Water is an inelastic fluid. It cannot be compressed or expanded. So water that is lost from one place must reappear in another. That cubic centimeter of water: How is it stretched and skewed, tangled and mixed as it leaves my creek and re-joins the sea?

I chuck in a dried-out seedpod and estimate the speed at which it floats away. I look for something else, something that will slowly sink, that will allow me to compare the surface speed with the speed of the deep water. I notice that the seaweed trailing from the jetties flutters at intermediate depths. I tear off a few shreds and plunge them into the water, releasing them at various depths. So far as I can tell, they all move off at the same rate. It is as if the whole of the sea is being pulled away.

The continued fall of the water has revealed more mud. Slopes that appeared to be gently shelving an hour ago are now revealed to drop off more steeply in soft, sticky cliffs. The miniature topography has a kind of sublimity—Norfolk's grand canyon. I find a patch of mud sharply illuminated by the still-slanting sun. The texture of the surface is extremely fine, with areas that gleam like chrome. There is some shade in deeper hollows that look as if they might have been made by children's footprints on the previous tide. Here and there, a

late drip begins its long journey seaward, producing a tiny glaucous wave front, like a contact lens sliding down a wet basin.

The apparently inanimate mud is chirruping vigorously now like sparrows in the bushes. It is the color of pâté on the surface. But immediately underneath, and for as far down as I can slip my hand, it is a fine-grained blue black. What makes it so? And why is it different on top? More questions. The sound is coming from little holes that lie in all directions, bubbling away like fumaroles on the moonlike surface. I look closely until I see signs of movement at one of the holes, and then pick up a handful of the mud surrounding it and sift through it, to find a beautiful millipede-like creature about 5 centimeters in length. It is lime green in color with a fine dorsal stripe of black. It tapers at the tail, which is red. I will discover later that it is an estuary rag worm. I thought it might be so named because it calls to mind old rag mops, which were made by tying together strips of random colored cloth, but in fact it is because of the creature's habit of hanging limply when picked up. It is a surprise to find such a phantasmagoric animal—its Latin name is *Hediste diversicolor*—in this almost colorless environment. I wonder what evolutionary advantage its color confers.

My tributary is almost empty now, its hard bed of pebbles and shells in the deepest part visible through the remaining inch or two of water that now runs discernibly downhill in its effort to drain the marshes behind—a task it will never have time to accomplish before the next tide.

A CAPRICIOUS POWER

Following my next tide reading, I have time to observe the flora. What grows varies vertically, according to the length of time in or out of the salt water to which each plant is adapted. This is what ecologists call "zonation," the tendency for organisms to inhabit favored biogeographical zones. All organisms, whether plants or animals, are subject to zonation, but this usually occurs over huge areas of land and has fuzzy boundaries, so the sense of a zone is seldom felt. The black-headed gull that I saw paddling and feeding, for example, has a range covering more than 3 percent of the earth's surface, which includes habitats from oceans to farmland across all of Europe, the northwest coast of North America, and much of Asia and Africa north of the equator.

But here, between the high- and low-water marks, the zones change abruptly. One sharply delineated biological border may lie just meters, or even centimeters, from the next, with each narrow ribbon stretching horizontally for as far as I can see along the ragged edge of the land. Starting with the tallest-growing plants on the highest ground, I make notes and take samples so that I can identify each one later. The most prominent species is an occasional bush that grows to about a meter high. It has small, succulent pods for leaves that graduate to red at the tips. The plant (I will learn later) is a *Suaeda* or sea blite. Also growing on ground that will be covered only a few times a year at the highest tides are a number of grasses. They vary in height, in the width and thickness of their blades, and in the number of their seed heads. Many may belong to a species of cordgrass introduced to prevent the erosion of estuarine mudflats. I find sea lavender too, quite unlike its terrestrial namesake.

Until the eighteenth century, when scientific cataloguing began in earnest, many people believed that every plant and animal on land had its analogue by or in the sea. The legacy of this belief is the many organisms that still carry "sea" as a prefix to their names, even though they often bear scant resemblance to their supposed counterparts. Thus it is that sea lavender flowers not on long, gray stems like garden lavender, but in low, desiccated clusters that rustle as I brush past them.

Much of the ground cover is made up of little bushes of sea purslane, whose sage-like leaves I understand are edible—favored by the new breed of Nordic chefs—although I have not yet dared try it. But I do take home for the table a little samphire, which is growing out of the drying mud pan at a slightly lower level. There are other plants: a large, spatulate-leafed plantain; the soft, gray, aromatic leaves of sea wormwood; the mauve and yellow flowers of the sea aster; and more that I cannot identify.

I had noticed earlier, happening to be in the same place one very high tide once before, how quantities of dead leaves and grass stems are borne along on the water, along with the foam bergs. I wonder how much vegetable matter is carried out to sea in this way, and what becomes of it, and how much marine detritus is carried in on the flood and deposited on the shore in exchange.

I am happy to see no plastic carried out by the tide, though. The high-water line is now characterized on this coast, as on many others, by a dotting of colorful plastic bottles, ripped fishing nets, and useless lengths of old rope, which shine out amid the rotting seaweed. This flotsam—or is it jetsam? I should discover the meaning of these curious words—is distributed along the shore by winds and tides and has often been carried great distances. The broken plastic beach spades and wrinkled party balloons I find washed up like diaphanous sea creatures have not come far. But occasionally I see a fragment of a fish crate from Norway or a bottle label in the Cyrillic alphabet. Clearly, it has a story to tell about our wastefulness and our blithe unconcern for our planet, but perhaps it also holds a clue to the movements of the tides themselves.

A little below the line where the sea purslane stops, the mud slope steepens suddenly. From having been so very flat over many acres— the flatness the sign of an environment dependent upon, and stabilized by, the presence of the sea—the land crumbles and dissolves. The first foot or so of this gradient, especially on south-facing banks, is often cracked mud, which shows that it is not covered by every tide. The samphire and other things I have seen growing on the flat do not grow here, but there are some new plants. One in particular catches my eye: it has multiple, tiny, artichoke-like heads, each containing many unopened yellow flowers.

The stratum immediately below is wetter and covered with a Prussian blue algal bloom. Then comes a band of apparently vacant mud, and after this the proper seaweeds. Dark wrack clings to the wooden bridge supports, straggling down-channel where the receding tide has left it. Here and there are fronds of a translucent green weed. I slip about wildly as I look around me. The diversity of flora is impressive now that I am looking closely. But it is the vertical ordering of all this variety that is truly remarkable; it is as if no plant dares to step out of line, fearing the tide that will shortly return to keep everything in order.

It is two o'clock. Slack water. The sun is hot, a last blast of summer that has come too late for the vacationers: I am alone on the marsh. The water is mirror smooth again, as it was at dawn, the light breeze unable to reach down below the level of the marsh to where the water has fallen. A last trickle—small but fast—carries some silt with it into the main channel, which spreads as a trailing skin across the surface of the slower water.

At the very bottom of the tide, I notice that the mud changes color again. A new stratum appears with a mustard-yellow bloom. Gulls have left their prints—including, I am surprised to find, occasional ones where, like me, they have lost their footing.

The water remains slack for a long time after low water, if I take low water as the halfway point between the preceding and following high-tide times. I am not sure if I ought do this, however, and my tide table is no help, since it gives only high-water times.

Suddenly, though, I notice something odd. Broad ripples are advancing up the flat-calm channel, against the still-running dregs of the ebb. New water is advancing over top of the old—a complex flow I had not counted on. Can it be that the tide is both going out and coming in at the same time? The ripples are gone a few moments later. Was it an illusion, a hallucination brought on by too much sun? Surely the tide doesn't tease, but moves with majestic, determined honesty.

By four, I am beginning to feel impatient. Far from being at one with the Zen of it all, I am wondering whether the tide will ever return. It is now more than three hours past the moment midway between the two highs. What can explain this apparent irregularity? Is the tide not a well-oiled machine like a piston pumping in the sort of oscillation mathematically described by the form known as a sine wave, which is indeed the very shape of an unbroken ocean wave? Is this disturbance something that happens everywhere, or just here? Does it happen on every tide here, or just on some? What perturbation of the cosmic clockwork is this?

The smooth passage of the moon and sun across the sky should surely produce a correspondingly even pulsation of the sea level. Why this irregularity? Why the delay? When I have the data from the entire tidal cycle, I will plot each point on a graph. It will show how the water really moves between succeeding high tides. The gradient of the line during the ebb and the following flood will reveal at a glance the speed at which the sea has emptied and then replenished my inlet.

Half an hour later, it is clear that the flood has begun in earnest. It does seem late, though; a queue of people has appeared waiting for a boat trip to see the seals that bask on the sands nearby, but the boats are still stuck fast in the mud. On this north-facing coast, a southerly wind can hold the tide back at sea, but the wind today is negligible. It is another mystery.

On a bend in the channel, the tide is running in fastest in a band that lies a little way in from the outside edge. Tendrils of green weed speed past faster than I would be able to swim or perhaps even walk. The *îles flottantes* are back too. Again, it is clear that not all the water runs at the same rate. Close to the edge, the tide runs more slowly, and the scum accumulates into a film that makes slow-motion Hoku-

sai wave shapes as it sweeps past me. Boats refloat with guttural bub-
blings from below. I watch the tide fill a gull print on the sloping mud
bank; it rises a centimeter in just ten seconds. The creeks are refilling
with what looks like unstirred coffee, there is so much sediment carried in—new organisms that will make this their habitat, food and
nutrients brought by nature's unfailing thirteen-hour delivery service,
new minerals that will gradually alter this inconstant landscape. New
debris that has settled on the mud during the hours of low tide is
swept in too—white feathers of gulls and egrets.

Land is being covered, and with it a landscape as well. The landscape will be back with the next ebb, but the actual land will be
changed, made anew, replenished with a new deposition of sediment
and nutrients. "No man ever steps in the same river twice," according to Heraclitus. The Greek philosopher was right, of course. Water
flows in only one direction in a river, and so it is constantly renewed at
any given point. But the same is true of the tidal sea, where the water
is always on the move, and it is never the same water reassembling on
one tide that was there at the last. In places, the tide adds; in others,
it takes away. No part of the land it touches will be quite the same
again either.

The flow seems deliberately directed now, although the water is
not looking for a destination. The tide is strong and steady in front of
me, and it is equally strong and steady both upstream and down: the
water is not heading to one place; it is filling space, or rather volume.
It is coming up even as it appears to be going along. At its fastest, the
flooding tide again creates eddies around obstacles in its path. I can
see that the water is actually maintained at a higher level nearer the
sea than it is where the channel snakes inland. The sea is sloping with
the sheer mass of water trying to shove its way forward, to make it on
time, to meet its unmissable appointment. If the water were to come
in much faster, I reckon, it would trip over itself; it would form the
kind of wave called a tidal bore.

In my little creek, the flooding tide has swung the moored wreck around so that it appears ready to go to sea. The water at the edges often seems lost or confused. Now I am taking readings every three minutes, the flood is so rapid: at 16:49, 0.20 meter; 16:52, 0.30 meter; 16:55, 0.39 meter; 17:00, 0.52 meter; 17:05, 0.61 meter; and so on. Before long the tide has risen far enough to meet the wind blowing across the top of the marshes. The breeze is getting up for the first time in the day, bringing with it a slight haze. I hear the sonar pings of redshank as the bird calls resume after the torpor of the afternoon, and the promise of a cooler evening is felt on the air. For the first time in the day, too, the wind is against the direction of the tide. Friction between the air and the water kicks up short wavelets. Out at sea, these characteristic steep waves are used by sailors to diagnose the actual state of the tide more reliably than printed tables and atlases do. They also create some of the most dangerous seas known to mariners.

Now I can begin to appreciate the astonishing phenomenon that is going on around me. The transport of so much water—such volume, such mass—at a cosmic click of the fingers. I have not felt an even breathing as the sun and moon pull on the earth's oceans; I have experienced a roller-coaster ride on which, although I have been duly returned to the point from which I left, the dips and rises have come with unexpected timing, as if on a whim. It is a capricious kind of power I have felt.

18:30, 1.83 METERS. As my day nears its end, I feel that I have at last sensed a different rhythm of things. I can appreciate twelve hours and forty-odd minutes, the time from trough to trough or peak to peak of the tidal cycle anywhere in the world, as one of the earth's fundamental units of time, to rank alongside the day and the lunar month and the year. But what I cannot appreciate simply from my day on the shore is why it is this particular length of time—so close to, and yet stubbornly always a bit greater than, half a day—that counts, and what relation, if any, this period bears to other celestial units of

time. I have a sense, too, as a man guns his outboard past me to race out to his fishing boat moored in the harbor, of what artificially accelerated lives we lead. I have seen people go out onto the marshes with their dogs for a walk and come back seemingly moments later. I have seen picnic boat trips accomplished and apparently enjoyed in equally short order.

Later, I view a short film made by the artist Susi Arnott. Called *Estuary*, it records two full tidal cycles in the Camel River in Cornwall. Using time-lapse photography, Susi has compressed more than twenty-four hours into just twelve minutes. At this speed, new things become apparent that I have not been able to observe in real time. The sun and moon track purposefully across the sky. Seen like this, it is easily possible to imagine they are dragging the sea along behind them. People in boats zip about like flies. The boats lying at their moorings swing from side to side like so many pendulums as the tide ebbs past them. They swing at similar, but not identical rates, the difference governed, I suppose, by the length of their ropes or the shape of their hulls or the speed of the current—I cannot tell which. They continue in this frenzied oscillation as the water gets shallower right up until the minute they find themselves aground. A sandbar in the middle of the river dries out gradually, a ridged loop of newborn land appearing like a fingerprint as the escaping tide finds its runnels and slithers out to sea. The scene is not exactly replayed in reverse when the flood returns. Instead, the new tide sweeps boldly across this now dry sand in a single front, probing ahead with watery fingers. At the turn of the tide, the breeze causes all the boats still floating to pirouette at once like a chorus line. Their mooring ropes stretch out once more, now straining upstream, and the frantic wiggling motions recommence. It is oddly mesmerizing. I have watched the DVD play singly, but I notice there is also an option to view it on a loop. I can see why.

In my estuary, the mud is now fully covered once again. The water is lapping around the vegetation at my feet. I begin to feel very strange.

Perhaps it is the effect of the long day in the sun or being alone on the marshes, standing only a few inches above the level of the water, with everything, land and sea, stretching out equally flat all around me. Even knowing that the rising tide has come to a halt and will be going out during the next few hours, I feel an incipient fear of isolation, even of drowning, of being absorbed forever into this in-between world. What if, this time, the water doesn't stop?

The wind has died down again. At quarter past seven, the tide reaches the height where I found it in the morning. It is amazing that the water can have risen so far and so fast and now be utterly quiescent, as if nothing here has ever changed.

But it does not rest. The ebb begins immediately. Even though no movement of the water is yet visible, the boats have sensed the change and are swinging on their moorings. Fifteen minutes later, the tide has already fallen a fraction. The new cycle is under way.

2

BEYOND THE MICROMAREAL

PHILOSOPHER'S END

In my travels to coastal locations for this book, I have typically been confronted when I reach my destination by conscientiously placed municipal signs warning people of the dangers of the tide. So perhaps I should issue my own warning here that this chapter involves two quite unnecessary watery deaths of mythic proportions from long before any safety-conscious bureaucrat thought to intervene for the public good.

The town of Chalcis is not the tourist's Greece. It lies within easy commuting distance of Athens on the long island of Euboea at the narrowest point of the channel that divides it from the mainland, which is here named Boeotia. Hidden behind waterfront apartment blocks, the old town occupies a rocky promontory on Euboea, while a new town has grown up on the opposite bank of what is now called the Euripus Strait.

The channel is extremely narrow at this point, even though the landmasses on either side are huge. The banks are in fact less than 200 feet apart, and there has been a bridge of one sort or another here since antiquity. Sadly, a fine multiarched stone bridge with a crenellated gatehouse that endured through centuries of Byzantine and

Venetian rule has been supplanted by a functional steel bridge that slides aside to allow boats to pass.

It was to this town that Aristotle, not long turned sixty, retired from the hectic round of intellectual life in Athens. He may have stopped writing and teaching, but he had not stopped observing and thinking, and one of the things he noticed was the way the water flowed through the strait. It is unusual for the waters of the Mediterranean Sea to move noticeably anywhere, but here they rushed back and forth in a flagrant manner that demanded an explanation.

The Euripus current runs now in one direction, now in the other, at speeds of up to 3 meters per second—too fast to swim against, often too fast for a boat. According to legend, it turns not like an ocean tide, but capriciously, at irregular intervals, seven times a day. Indeed, the strait is named Euripus after the Greek word for "weathercock" because of this erratic behavior; "Euripus-heads" were people who were unable to make up their minds. The myth of the place meant that *euripus* spread as a generic word for any fast-flowing channel.

Aristotle did not believe Plato's ideas that tides were the remnant eddies of the outflow of rivers into the sea or, more fancifully, the bodily fluids of the animal earth. Other Greek philosophers effectively declined to address the problem, attributing the tides to the gods and leaving it at that.

At Euripus, Aristotle studied the currents and deduced that they were unrelated to the wind, but he got no further toward a solution. In 322 BCE, disgusted by his lack of progress, he threw himself into the churning waters to be drowned. This, at least, is the polished version of the story developed by late-antique historians. In the most melodramatic telling, Aristotle leaped into the current, exclaiming, *Si quidem ego non capio te, tu capies me* ("If I cannot grasp you, then you must take me"), although most scholars today find no reason to believe that Aristotle died of anything other than natural causes.

As time passed, so the stories—and the turnings of the tide— multiplied. Some believed that the current turned not seven times in a day, but fourteen. Some concluded that the Euripus Strait behaved in a regular fashion for most of each lunar month, with only a few days during which the water went haywire, although they failed to note the phase of the moon in which the change occurred. The revival of interest in classical scholarship prompted Renaissance natural philosophers to look again. The English writer and physician Sir Thomas Browne could not believe that one of his scientific heroes would have committed suicide as the legend claimed, and he tried to debunk the myth in his 1646 catalogue of foolish beliefs, *Pseudodoxia Epidemica*. After examining a range of classical writers' opinions on the matter, Browne relates the visit of one Monsieur Duloir sometime in the 1620s, who, being of an empirical frame of mind, hired a boatman to take him to a suitable vantage point for the day in order to make direct observations. He found that the tide turned just four times a day, as one would expect in most tidal waters. Perhaps, though, he simply happened to be there at the wrong time of the month to witness anything more extraordinary.

Aristotle's reputation is rescued by modern science. In 1997, Mikis Tsimplis, a Greek oceanographer then working at the Proudman Oceanographic Laboratory in Liverpool, decided to reexamine the problem. Tsimplis tells me that one of his chief interests is "marine extremes," and he has long been intrigued by the Euripus Strait, visiting it on several occasions. In his analysis, he used data gathered from a string of tide gauges located north and south of the narrows to model the tidal pattern. He found that the range between high and low tide was four times greater on the north side than on the south, and that the resulting gradient of the sea level at various times in the tidal cycle was sufficient to generate large currents. In addition, there is a lag between the times of high tide north and south of the strait because of how long it takes each tide to travel around the island of Euboea. This effect adds to the height difference between the two sides and causes a fast current to flow one way or the other.

But the picture is even more complicated. The channels both north and south of the Euripus Strait are constricted by further narrows, each about 40 miles distant from Chalcis. Each of the two, almost enclosed bodies of water on either side of the strait, as well as being subject to the gravitational attraction of the moon and sun, has its own natural period of oscillation, like water in a bath. Oscillations can easily be set in motion by sharp changes in the local weather. This has the effect of enhancing the tides here compared to those seen in the Mediterranean generally. On the north side, this period is close to four oscillations a day, but on the south side it is more like six. Because these oscillation periods are unequal, the water gradient through the narrows is sometimes greatest at times other than those that would be expected from the action of lunar and solar forces alone. Tsimplis's analysis finds nothing out of the ordinary on the north side of the channel—which is perhaps where Monsieur Duloir's boatman took him—but on the south side he finds a pattern of strong tidal flow up to six times a day under certain conditions.

The effect may have been even greater and more complicated in Aristotle's day, as the strait was widened a little when the new bridge was built, causing the current to run more slowly than it may once have done, though still twice as fast as a good walking pace. Nevertheless, the flow under the bridge at Chalcis remains powerful and erratic enough to serve as a general notice that, if the local conditions are suited, the tide almost anywhere can give rise to strikingly peculiar effects, even when its intrinsic resources are modest. After all, the tidal range at Chalcis is never greater than 35 centimeters. The following chapters will lead us to many more such places.

MEDITERRANEAN TIDES

The greatest tidal range in the Mediterranean occurs at Sfax in modern Tunisia, but even here the tide never rises or falls by much more than a meter and a half. Other places where the tide is significant

include the north end of the Adriatic Sea, affecting Venice and Trieste, and at the Pillars of Hercules, the westernmost narrows that
lead out into the Atlantic Ocean. Here, the tidal range is about 50
centimeters, whereas not far out along the Atlantic coasts it is 3 meters
or more. In some locations within the Mediterranean, such as around
the Balearic Islands and in the Aegean Sea, there is hardly any tide
at all.

This does not mean that the water never moves, however. Several places in the Mediterranean occasionally experience the *rissaga*,
a kind of miniature tsunami generated when severe weather produces
a sudden change in the pressure of the atmosphere bearing on the
sea surface, which in turn causes a substantial "tide" to rise and fall
again within the space of perhaps no more than ten minutes. Known
more generally as seiches, these events were first observed in 1890
on Lake Geneva, where there is, of course, no ocean tide. The port
of Ciutadella on Minorca is especially vulnerable because, as is now
understood, the size and shape of its harbor inlet give it a natural
period of oscillation that happens to resonate with the typical weather
disturbance. On June 15, 2006, an exceptional *rissaga* caused the sea
level to rise by more than 4 meters, producing a surge wave that sank
thirty-five boats moored in the harbor and damaged many others.
The right atmospheric conditions tend to occur in June, when the sea
is still relatively cold and warm air blows north from Africa. For this
reason, the *rissaga* has come to be associated with the Feast of Saint
John on June 24, although the idea of a celebration based around this
natural surge is probably much older. "My theory is that it all comes
from people seeing that phenomenon happen, probably making the
harbor useless," says Graham Giese, emeritus professor of coastal
oceanography at Woods Hole Oceanographic Institution in Massachusetts. "That must have been a huge thing. It's still a huge thing."

Notwithstanding such localized phenomena, the Mediterranean
Sea is what scientists—especially marine biologists, for whom this
status has far-reaching implications—term a "micromareal" region.

Micromareal means that the Mediterranean hardly qualifies as a proper *mare*, like the oceans or even an otherwise unremarkable body of water such as the North Sea, which are macromareal, having generally substantial tides. It is this micromareal regime of generally insignificant tides that has created the Mediterranean of benign seas, stable beaches, and easy harbors that so many of us unthinkingly enjoy today.

SCYLLA AND CHARYBDIS

The Euripus Strait is not the only Mediterranean location that the ancients regarded as noteworthy. In Book XII of *The Odyssey*, the homeward-bound Odysseus pauses on the island of Aeaea, where the goddess Circe warns him and his oarsmen of dangers that await them: the bewitching Sirens, and then the rocks of Scylla, below which "dread Charybdis sucks the dark waters down" and spews them out three times a day—malevolent hazards "whom no sailors pass unscathed."

Noting Circe's instructions, Odysseus gives orders for the voyage to continue. His crew row safely past the Sirens with their ears stopped with wax. Almost immediately, though, they see "a cloud of spume ahead and a raging surf" and hear the thunder of breakers. Odysseus directs his men to steer close by Scylla's rocks to avoid the still more perilous whirlpool.

> Thus we sailed up the straits, wailing in terror, for on one side we had Scylla, and on the other the awesome Charybdis sucked down salt water in her dreadful way. When she vomited it up, she was stirred to her depths and seethed over like a cauldron on a blazing fire; and the spray she flung up rained down on the tops of the crags at either side. But when she swallowed the salt water down, the whole interior of her

vortex was exposed, the rocks re-echoed to her fearful roar,
and the dark blue sands of the sea-bed were exposed.

Transfixed by "Charybdis as the quarter from which we looked for disaster," they sail too close to Scylla, who plucks six crewmen from the boat and devours them on her rocky shore. Odysseus escapes with his remaining men and eventually reaches the comparative safety of Hyperion's Island of the Sun.

But where is this most famous of whirlpools, this tide-made terror? Certainly, its location cannot be deduced from Homer's phantasmagoric tale. The great storyteller is famously vague on geography. However, Charybdis also appears in the writings of the Roman Virgil, who drew on the more accurate description of it by the historian Sallust, and in various other classical accounts. The Roman geographer Strabo, for instance, though he believed that the tides were caused by movements of the seabed, was clear that the thing is to be found in the narrow Strait of Messina between the island of Sicily and the Italian mainland. But modern commentaries on the classical authors throw new doubt on the matter. For example, in his commentary on Sallust's *Histories*, Patrick McGushin notes, "The name was identified with the Straits of Messina in which, in fact, no such phenomenon exists."

I wonder whether, in fact, the Charybdis of classical mythology wasn't genuinely deserving of its fearsome reputation but in the centuries since, some alteration of the coastline or seabed has caused it to become less spectacular. Whether it exists now, or existed then, or was only ever a pale imitation of the man-eating horror described in the stories, we are left today with the expression "between Scylla and Charybdis" to describe our predicament when we, like Odysseus, are forced to negotiate a passage between two possibly contrasting, but equally hazardous, alternatives. There are, too, this idiom's more generic derivatives, "between a rock and a hard place" and "between the devil and the deep blue sea," for although Scylla is a she-fiend and not a male devil, the sea is indeed notably deep as well as typically blue in the Messina Strait.

I do not believe McGushin's contempt is justified. Perhaps he does not understand that tidal places are temporal as well as spatial occurrences, coming into existence only during those hours when the current is running sufficiently strongly one way or the other. (He doubts, too, that the waves there can make the barking noise ascribed to them by Sallust, although I find it eminently believable that the sharp chop whipped up when the fast current swirls against the wind would slap and yelp like stray dogs in a Sicilian back street.)

I am tempted to visit and see for myself. The rocky headland of Scilla is clearly labeled on modern maps, jutting out from the Calabrian coast, with a lighthouse to warn ships of its fabled dangers. But Charybdis—or Cariddi, as it is called locally—does not appear. It is neither on the maps—understandably, perhaps, because it is a liquid, changing feature of the seascape and not a solid terrain fixture—nor even in the tourist guides. Only Michelin manages a grudging reference to *The Odyssey*.

Yet I have my reasons to hesitate. This is a path I have trodden before. Once, fired by lurid memories of Aeneas's journey into the underworld in Book VI of *The Aeneid*, part of which had been the set text in my Latin "O" level, I went in search of Avernus, the stinking lake of which Pliny wrote that even birds flying toward it die, which appears in Virgil's epic tale as the dangerous threshold where Aeneas seeks instruction from the Sibyl of Cumae. I found it; it was more of a pond, surrounded by reeds, with dull landscape around, and a metallic-blue surface of unreadable depth. It is true I saw no birds, but then neither would I expect to on a municipal boating lake, which was what Virgil's lake innocuously resembled. Others have found Charybdis just as underwhelming. Peter Stothard, for example, who clearly suffers from a more extreme version of Classical Quest Syndrome than I, writes in his book *On the Spartacus Road* that "Scilla and Charybdis have been a disappointment for as long as tourists have taken any kind of classical trail." And Hilaire Belloc is downright scornful: "I have seen Charybdis—piffling little thing." So I shall forsake authenticity on this occasion—I have, in any case, plans to take you to grander and more terrifying tidal sites far beyond the tame Mediterranean and the ken of classical authors—and let another fable stand in for observation.

AN EXPERIMENTAL EMPEROR

Homeric legend notwithstanding, the story that I find best illustrates the mystery of Charybdis, and the urge to comprehend it in a scientific way, comes from the reign of Frederick II, the Holy Roman emperor and king of Sicily in the thirteenth century. The red-haired Frederick was a rare figure among medieval rulers, being both efficient and enlightened in the governance of his far-flung kingdom. Combining German, Norman, and Sicilian blood, he cherished Sicily as a meeting point of Greek, Latin, Jewish, and Arabic culture. He founded the university at Naples, but the court in Palermo was the intellectual center of his world. Here, he pursued philosophy, mathematics, and natural science, and displayed the Epicurean tendencies that led him into repeated conflict with the church. He was noted by his subjects for his tolerance and also his stubbornness.

But to us today, it is Frederick's rational and inquiring mind that stands out. He employed two scientific advisers—one from Scotland, the other from Egypt—who produced manuals on hygiene and veterinary science, although Frederick's intellect was undoubtedly the greatest of all. He himself wrote treatises on horses, birds, and falconry. He banned trial by ordeal because he could see it was a method of justice that favored not the innocent, but simply the fittest. He instituted the legal separation between physicians and apothecaries to ensure that doctors could not profit by selling prescriptions—a system that we still abide by today.

Frederick also understood the merit of empirical observation and the power of experiment. He documented his own results too—for example, refuting the myth that barnacle geese hatch from a particular species of barnacle, which they fancifully resemble, and demonstrating, by sealing their eyes, that vultures are able to find their food by scent alone. Overall, his method is remarkably scientific: statements are supported by facts, not opinion, and if information is incomplete, the conclusion is left open. He asks the scientist's question—how?—albeit sometimes of impossible things. How many heavens are there? How does God sit on his throne? But also, How large is the earth, and how is it that some waters are salt and some are sweet?

The record of some of the more bizarre scientific experiments that Frederick ordered comes from the contemporary Franciscan friar and chronicler Salimbene di Adam, who seems to have been impressed as a fifteen-year-old boy when the king and his menagerie of exotic African animals visited his home city of Parma. In adulthood, Salimbene was a more qualified admirer of the sinful king, inclined to forgive his master's excesses because he had so many enemies and because he apparently had a fine sense of humor. Frederick, according to Salimbene, was "a comely man, and well-formed, but of middle stature." He spoke many languages, and "if he had been rightly Catholic, and had loved God and His Church, he would have had few emperors his equals in the world."

Salimbene gives a good impression of an impartial biographer here, but it is hard to know whether to trust him when it comes to recounting some of Frederick's wilder exploits. The chronicler finds too many regal "eccentricities" to relate and restricts himself to just seven of the more reprehensible examples, including an occasion when the king ordered the thumbs to be cut off a lawyer who had written his name differently from the way he wanted, putting "Fredericus" instead of "Fridericus." In a more investigative, though hardly more humane, spirit, Frederick designed an experiment to establish what was humankind's "original" language by arranging to have raised from birth a number of infants who would be nursed without hearing speech or even baby talk. Would it be Hebrew ("the first language") or Greek, Latin, or Arabic they would speak? Or would it be the language of the parents from whom the unlucky infants had been parted? The king never found his answer, as the babies died before they could speak. On another occasion, he fed two men and sent one off to sleep and the other to hunt, and then had them both disemboweled in his presence to see which of them had digested his food better. Another experiment sought, by shutting a man in a barrel until he died, to establish whether, as the skeptical king thought, the soul perishes along with the body at death.

But it was the king's "fourth eccentricity" that concerned the mysteries of the deep in the wild waters of the Messina Strait. The story centers on a young man of Messina who was such a fine swimmer that he was known as Colapesce, or "Nicolà the Fish." Wishing to know what it was that gave rise to the whirlpool of Charybdis, Frederick instructed Nicolà to dive down into the churning waters and report back what he saw. After several dives by the reluctant Nicolà, the king was still unconvinced that his subject had investigated properly:

> And Frederick, wishing to know the definite truth if Nicola had really been to the bottom before resurfacing or not, threw in his golden cup where he thought it to be the deepest. And Nicola went down, found it and brought it up; and the emperor was amazed. With the emperor wishing to send him down again, Nicola told him: "No way should you want to send me in there, because at the bottom it's so rough that if you do, I won't make it back." Notwithstanding, the Emperor demanded he dive, and Nicola did not reappear because there he died.

From the knowledge of his Franciscan brethren in Messina, Salimbene adds, "At the bottom of the sea are great fish at time of storms and there are rocks and many broken ships, as this same Nicola has told. He was able to tell Frederick what is written in Jonah 2: You hurled me into the depths, into the very heart of the seas, and the currents swirled about me; all your waves and breakers swept over me."

Salimbene makes clear what the older mythology tends to obscure, which is that Charybdis is a transient and variable feature of the sea: "There is *sometimes* a violent current which forms violent eddies," he says (my italics). The famous whirlpool—or, more likely, a series of ever-changing whirlpools—may arise at the turn of the tide when currents flowing in opposing directions are forced together. This is because the two currents behave in different ways. The incoming tide enters the strait, which becomes narrower, beginning by flowing in evenly across the full width of the channel and then speeding up as it is squeezed into the narrows. When the ebb tide runs out from these same narrows, on the other hand, it naturally continues in a straight path, like water pouring from a tap, instead of spreading across the widening channel. The old and new tides therefore pass by one another, causing turbulence where they become caught up, which may, if the current is fast enough, develop into whirlpools.

Normal tides rise and fall twice a day, as we know, so why does Homer clearly speak of Charybdis arising and subsiding three times a day? In fact, despite all the action off the coast, the tide at Messina rises and falls hardly at all; seawater flows this way and that through the strait, but the sea level remains almost constant. In other words, the port of Messina happens to lie at what oceanographers call an amphidromic point, a point around which tidal movements revolve, but where the level of the water does not change, like the calm center of a storm or the nodal midpoint on a vibrating violin string. "Amphidromic" means "running around," and it so happens that the time the tide takes to run around the island of Sicily is just different enough from the period of a normal tidal cycle to introduce complications to the expected pattern. This means that the tides in the Messina Strait, like those in the Euripus Strait, fill at different times from the Tyrrhenian Sea to the north and the Ionian Sea to the south.

In addition, the seawater is strongly layered, with a colder, denser, saltier layer underlying a warm surface layer, and these layers also tend to travel in opposite directions. A further complicating factor may be the wind, which tends to gather strength during the course of a hot day, and which, in a sea surrounded by mountains, may blow from unexpected directions. It is probably the combination of the wind against the tide that gives rise to the breakers that Odysseus's men heard yelping like dogs.

Salimbene is notably correct that some of the creatures found in the region of the Charybdis whirlpool are unusually large. As in many places where the tides conspire to create unusual undersea conditions, marine biodiversity is unusually high. Cold, salty water thrust up by the rocky sill lying across the strait raises nutrients up from the depths, which sustain a food chain from plankton and algae to crustaceans and large fish. It also brings alien "abyssal" fauna such as viperfish and bioluminescent lantern fish closer to the surface and to fishermen's nets. Fish passing through the strait, in turn, help to support a diverse bird population, making the strait an important stop on many species' migration routes. A few of these marine organisms fully justify the title of sea monster, such as the gorgonocephalid basket star with its baroque branching arms and the anglerfish with its gaping maw and ominously trailing lures. Why, I wonder, do creatures that live at great depths look so much odder than those that live near the surface? Is it that survival in conditions of almost no light and extraordinarily high pressure drives or frees them to assume horrendous shapes? Or is it just our familiarity with shallow-living fish that blinds us to the peculiar natural beauty of these rare ambassadors from the deep?

Either way, it makes Messina the perfect location for the Italian Institute for the Coastal Marine Environment, where the fields of study range from ecology to physical oceanography. In fact, Germans were the first to subject the peculiar fauna of these waters to proper scientific scrutiny, in the nineteenth century, drawn to the area not only by nature's exotic promise but by the sun and perhaps the romance of Goethe's *Italian Journey*. Today, projects of the Messina institute include mapping fish resources and evaluating the prospects for extracting power from the furious currents.

The peculiar pattern of the currents in the Strait of Messina, which the Greek geographer Eratosthenes in the third century BCE likened to full Atlantic tides, has recently been verified. Mauro Federico of the University of Messina found that the current reversed every six hours and eight minutes, owing to the alternation in influence of the flow from the Tyrrhenian and the Ionian Seas created by the presence of Sicily itself.

It has taken more than two thousand years to unravel these classical mysteries from the tranquil waters of the Mediterranean Sea. How many similar phenomena await scientific explanation in the world's oceans? How many even await discovery?

ALEXANDER AND PYTHEAS

A few Greeks did travel beyond the Mediterranean to find themselves on more challenging shores. Herodotus may have been the first. He had seen large tides on the Red Sea in about 450 BCE, but it was an isolated observation, scarcely to be believed back in Greece.

In 334 BCE, Aristotle's star pupil, the twenty-two-year-old Prince Alexander of Macedon, set out from Greece with an army of more than forty thousand soldiers to defeat the Persians and conquer Egypt, thereby expanding the Greek Empire far beyond its European base. The fighting was hard, and Alexander himself was wounded more than once. Eventually, with his troops at the point of mutiny, and the Himalayan mountains blocking further progress, he at last turned south for home. By the summer of 325 BCE, they stood on the banks of the Indus River at Pattala, near modern Hyderabad in Pakistan. While most of the army set up camp on the riverbank, Alexander led some ships down the river to establish how far they were from the sea. He took no local pilot with him, but he learned when he questioned some natives completely ignorant of the sea that the river turned salty two days' passage downstream. After two days more of rowing, the Macedonians moored up on an island in order to gather supplies, when they suddenly found the river around them flowing upstream. The rushing water smelled and tasted of the sea. Perhaps they had already begun to congratulate themselves that they were close to being able to sail for home, when a large wave surged up the river toward them. "At first, the river-current was arrested; then, driven with increasing violence, it ran backwards with greater force than that of torrents rushing on a down-hill course," wrote the Roman historian Quintus Curtius Rufus three centuries later. Wave after wave ran unbroken from bank to bank and flooded over the dry fields. As the waves passed, they smashed the boats against one another; "one might have thought it was not a single fleet that was on the water, but the ships of two navies starting a sea-battle," as Curtius Rufus put it. "Then, suddenly, they were struck with fresh terror": the water level dropped back again as suddenly as it had risen, leaving the boats high and dry, and army gear scattered among gulping unfamiliar sea creatures. "They could scarcely believe their eyes as they looked at what was happening— shipwreck on dry land and the sea in a river!"

That night, Alexander sent a detachment of cavalry downriver to give warning in case there were to be any recurrence of this unprecedented phenomenon. It was a wise precaution as it turned out. The men came racing back to the main camp ahead of a second wave, which surged up the river some twelve hours after the first. The wave was smaller this time, but sufficient to enable the stranded ships to be refloated, and "the banks rang with cheers from the soldiers and sailors as they welcomed their unexpected rescue with exuberant joy."

Alexander was not to know it, but he and his men had been caught up in one of the most dramatic and destructive of all tidal phenomena, a tidal bore. (A single wave we might be inclined to ascribe to a tsunami, perhaps raised by an undersea earthquake, but the recurrence of the wave, just over twelve hours after the first, as noted by Alexander's historians, confirms beyond doubt that this was a tidal phenomenon.) The Greek army continued its exploration down to the coast of the Arabian Sea, made a sacrifice to Poseidon, founded one more settlement called Alexandria—as they had done a dozen times before—and headed for home.

LESS CELEBRATED IN HISTORY, but far more significant for our lasting knowledge of the tides, was the voyage taken in the opposite direction by Pytheas of Marseille at around the same time that Alexander the Great was nearly carried off by the Indus bore. Stories of Hyperborea—the cold land beyond the north wind—were already circulating in the Greek Mediterranean, repeated by Herodotus, among others. But these tales had no real substance. Pytheas was to change that. He traveled overland to the Gironde River before taking a ship to Brittany, and then on to Cornwall and up through the Irish Sea, perhaps as far as Orkney and Shetland. Legend even has it that he sailed above the Arctic Circle to a land that he called Thule.

Pytheas's own account of the voyage is lost. The stories are contained in the much later writings of the geographer Strabo and Pliny the Elder. Nevertheless, as the noted archaeologist Barry Cunliffe has observed, "The sea and its behaviour evidently excited Pytheas' enquiring mind and he is known to have referred to it several times." Sailing out of the Gironde estuary, he found himself "in a totally different world, experiencing tidal displacements of up to 15 m. in many places." As reported by these later writers, Pytheas describes places that may only be reached by tidal causeway—perhaps those now known as Mont-Saint-Michel in Normandy and St. Michael's Mount in Cornwall—as well as places where, Pliny assures us, "tidal waters swell to a height of 80 cubits." This impossibly large figure—120 feet, three times greater than any tide Pytheas could have witnessed— reflects not only Pliny's lively storytelling but also Mediterranean people's sheer unfamiliarity with tides and just how large they should expect them to be. Another Greek, the philosopher Aetius, recorded Pytheas's observation that high- and low-tide times seemed to cor- relate with the fullness and faintness of the moon, although this, as Cunliffe points out, may be a confused reference to the difference in the average height of spring tides (which occur at full and new moon) and neap tides (which occur at half moon) rather than the twice-daily highs and lows.

Aside from Alexander the Great and Pytheas, it seems only a few other intrepid souls caught sight of the greater tides that lay beyond the Mediterranean. The polymath scientist Posidonius described the tides at Cádiz, but his original writings were lost, like those of Pyth- eas (destroyed, in his case, by the fire that razed the ancient library of Alexandria), and survive only in the reportage of Strabo. The Baby- lonian astronomer Seleucus of Seleucia, who was likely born on the shores of the Red Sea, first noted that the height of the high or low tide is not constant but alters gradually from one tide to the next in a way that seemed to be linked to the phases of the moon.

How much more quickly might Western civilization have developed a scientific comprehension of the tides if its cradle had lain not in the near-tideless Mediterranean, but along more changing shores. And how much more deeply inculcated in our own lives now might an understanding of the tides be if this comprehension had been part of our earliest understanding of the natural world. Perhaps it is this accident of Western history that is the deep seat of our igno- rance about the tides.

We happily attribute the beginnings of our mathematics, geome- try, architecture, physics, mechanics, astronomy, and natural history to the Greeks. It seems hardly believable that they would not have had something percipient to say about the tides if they had regarded them as significant at all. Yet for the Greeks, and the Romans after them, the modest oscillation of the tides was frequently lost altogether amid the larger effects of local winds and weather. It is unreasonable of us to expect them to have identified a regular pattern of movement of the sea. Even if they had been able to deduce a tentative connection between the moon and the tide, the Mediterranean would always be their benchmark. The huge tides that lay across the desert or through the Pillars of Hercules might have been taken not as a thumping confirmation of their theory, but as a completely different phenomenon of nature. Instead, the few tidal effects worth remarking were incorporated into myths and legends, as we have seen.

Biblical societies, largely sharing the same Mediterranean shores, were at least as ignorant of the tides as the Greeks. The ancient Egyptians, their land bordered again by the Mediterranean to the north, and by the Red Sea to the east, were also incurious.

What of those ancient cultures that were not so blessed as to become the cornerstone of Western civilization? The great majority of the world's tidal coastline lies outside Europe. Did not any of these civilizations learn something about the breathing oceans that lapped their shores? Some can be excused: the tides of Japan and many Pacific islands are negligible. The Mayans had an obsession with calendars, which included lunar phases in their reckoning, but understandably they did not connect these with the minuscule tides in the Gulf of Mexico. The Chinese made no significant breakthrough until the eleventh century, when the polymath Shen Gua suggested that the moon was a more important factor than the sun.

The best early knowledge came from the Harappan people of the Indus Valley more than four thousand years ago. In Gujarat, where the Arabian Sea brings large tides, archaeologists have uncovered evidence of a tidal dock dating from this time—a human-made basin that would fill with water at high tide linked to an estuary by a short canal with a gate to retain the water when the tide went out again. Centuries after they were aware of the practical utility of the tides, Indian societies attempted to explain them. The Samaveda, one of the Vedic scriptures, assembled around 1200 BCE, stated the link with the moon. The Hindu epic the *Mahabharata* made the explicit connection between spring tides and the full and new moons, and later the Puranas made the theoretical jump from believing that the oceans grew and shrank with the tides to the correct assumption that they maintained a fairly constant volume, with tides high and low in different places at the same time.

A WIDER DOMINION

DOVER, KENT

	LW	HW	LW	HW
August 27, 55 BCE		07:31		

In September of 54 BCE, when Julius Caesar's major invasion of Britain was under way, Marcus Tullius Cicero in Rome replied to a letter sent by his brother Quintus, who was with Caesar, making plain the apprehension the Romans felt about tidal shores. "How glad I was to get your letter from Britain! I was afraid of the ocean, afraid of the coast of the island. The other parts of the enterprise I do not underrate; but yet they inspire more hope than fear."

Accustomed to a sea level that altered very little, and to fast currents only in a few confined channels here and there, the Romans had reason to be fearful when they invaded Britain. There were practical considerations too. Triremes rowed by slaves might provide adequate power when becalmed in the Mediterranean, but if the wind dropped in Atlantic waters, then these vessels would be at the mercy of the tides.

Both of Caesar's landings in Britain, in 55 and then 54 BCE, were hampered by the strong and unfamiliar English Channel tides. "The Deified Caesar crossed over to the island twice," Strabo relates, "although he came back in haste, without accomplishing anything great or proceeding far into the island, not only on account of the quarrels that took place in the land of the Celti, among the barbarians and his own soldiers as well, but also on account of the fact that many of his ships had been lost at the time of the full moon, since the ebb-tides and the flood-tides got their increase at that time."

The exact point where Caesar and his troops set foot on British soil for the first time, in August of 55 BCE, is not clear from contemporary accounts and has become a topic of heated dispute among later historians. Finding the solution to this historical puzzle depends on a proper understanding of the tide as much as on the analysis of military intentions.

Caesar's fleet set sail from two ports in France early on the twenty-seventh of August, seven tides before the full moon, as he wrote in his firsthand account, *De bello gallico*. He had with him about eighty transport ships, enough for two legions, with additional ships due to bring supporting cavalry. The intended landing place was Dover, where there was a natural harbor. Arriving offshore a few hours later, however, the Romans saw the Britons lined on the cliff tops overlooking the port ready to defend their island; putting men ashore would be suicidal. Instead, Caesar tells us, he stood at anchor for five hours, until the middle of that afternoon.

It is in this interval of time that the tidal uncertainty arises. Caesar's account tantalizingly says that he eventually made landfall 7 miles away, "meeting both with wind and tide favourable at the same time." But 7 miles away in which direction? Unfortunately, there is no unambiguous description of the beachhead in Roman accounts of the invasion, and no archaeological trace has been found either. A pretty, stone memorial with a relief sculpture of Caesar's head placed at Walmer Green seems presumptuous in the circumstances.

IN 1862, two Victorian gentlemen took it upon themselves to resolve the matter once and for all. Thomas Lewin reckoned that the fleet would have been swept westward to Hythe, while the Reverend Edward Cardwell argued the opposite, saying that it would have been carried east around South Foreland toward Deal—both places more than 10 miles in their respective directions along the coast from Dover.

What drove these men in their eccentric quest—apart, that is, from an obvious excess of leisure hours? At stake was no less than certain knowledge of the site where civilization arrived on these islands—a matter of no small consequence for the high-minded builders of the British Empire. Very soon, the whole discussion was being referred up to the highest military and scientific authorities in the land, eventually reaching the desk of the Astronomer Royal himself.

The discussion began in earnest when the president of the Society of Antiquaries, Lord Stanhope, seeking an adjudication between Lewin and Cardwell, called upon the expertise of the Duke of Somerset, who was First Lord of the Admiralty, and so presumably knowledgeable about all things marine.

The time of high water on that morning of the twenty-seventh of August had been calculated at 07:31. By midafternoon, therefore, in Lewin's estimation, the tide would have been running out at full speed, and would have brought Caesar's fleet to shore somewhere near Hythe. His disputant Cardwell, however, had spoken with mariners in Folkestone who confided that the tides locally do not run as indicated in the Admiralty tide tables for the English Channel. Instead, an unmeasured inshore tide would have given Caesar's fleet slack water with a tide just beginning to run east—the Deal option. Cardwell had even interviewed the Folkestone ferry captain—"one person above all others at Dover on whose judgment reliance would be placed in a disputed question of this nature"—who would be sure to know the difference between the tide close to shore and the tide out at sea. The captain had confirmed that if Caesar had weighed anchor at low water, he would have been carried east and north around South Foreland, and not west as Lewin was thinking.

Rear-Admiral Lord Clarence Paget, secretary to the Admiralty, replied to Stanhope, confirming that the published Admiralty data were for the middle of the Channel. However, he offered to obtain entirely new measurements for the inshore waters, enclosing a chart for Stanhope to indicate where especially he would like readings to be taken. Stanhope duly marked the chart in pencil over an area from South Foreland to beyond Shakespeare Cliff, on either side of Dover, about a mile and a half out from the coast. Seven months later—it required four full lunar cycles in which to take the measurements, as well as time to write up the results—Paget wrote back enclosing the report of the Admiralty surveyor appointed to the task of data gathering. The inshore tide was found to turn four or five hours after high water, well after the main tide turned in mid-Channel. The surveyor added that he assumed it ebbed and flowed for nearly six and a half hours each way—a gross assumption that rather missed the crux of the problem, which is the consideration that the tides might run for unequal durations in the two directions.

In the end, Sir George Biddell Airy, the Astronomer Royal, ruled on the matter. Noting first that his illustrious predecessor, Edmond Halley, had looked into the times of the tides, but not the place, of Caesar's invasion, he went on to refute Cardwell's suggestion of a beachhead at Deal or Walmer as "absolutely impossible"—although, wisely no doubt, he refused to nominate an alternative landing point. Cardwell had gone astray misreading the scientific literature, taking a time of day (3:10 p.m.) for a duration of tidal flow (3 hours, 10 minutes). He had also fallen victim to a landlubber's misconception that if the tide is flowing away from a place, then the sea level there must be going down, which is often not, in fact, the case. The elucidation prompted Airy to an instructive reminiscence:

> When a young man, and before I had studied the theory of tide-waves, I used to visit the works of the new London Bridge; and on one occasion, standing on one of the central piers, I saw, to my great surprise, that, though the water had fallen more than a foot, the tidal current was still running rapidly up the river. I mentioned this to my late lamented friend, Mr. (afterwards Sir William) Cubitt, and he answered, "The water must flow upwards still, to produce high water in the upper reaches of the river." I have seldom learned so much from a few words as I did from this answer. But the law which I then learned is very rarely, almost never, known to water-side dwellers, whose ideas of high-water are usually connected merely with the cessation of flow. Perhaps there is sometimes a confused idea produced by the combination of the two.

For his part, Julius Caesar withdrew for the winter to prepare for a much larger invasion the following year, comprising five legions and cavalry that would be carried in new ships designed for easy beaching.

The Romans soon learned to master these difficult new seas. Tacitus writes in *Agricola* of the Roman fleet circumnavigating the island of Britain: "Nowhere has the sea a wider dominion . . . it has many currents running in every direction . . . it does not merely flow and ebb within the limits of the shore, but penetrates and winds far inland, and finds a home among hills and mountains as though in its own domain." Elsewhere, Tacitus refers to the invasion of the isle of Mona (Anglesey), where the Britons had retreated. By this time, around 60 CE, the Romans were clearly more familiar with tidal flow. The Menai Strait, which divides the island from mainland Wales, is known for its strong and treacherous tides. Undaunted, Agricola (Tacitus's father-in-law) cannily ordered flat-bottomed boats to be built that could be easily tided over the shallows rather than facing the currents in full spate.

These may have been the first occasions on which the Britons were bested in the tidal matters of their own shores. They were certainly not the last.

3

SHORES OF
IGNORANCE

AN EXPERIMENTAL KING

Picture the famous scene. That redoubtable Edwardian children's history of Britain, *Our Island Story*,* will help. Its author, H. E. Marshall, paints it well, complete with lively reported speech. Canute, the eleventh-century Danish king of England, is wise. His nobles are foolish flatterers. "Even the waves obey you," they assure him. The exasperated Canute sees that they will not be dissuaded from this idiotic view by argument alone. They need the evidence of their own eyes, and so he arranges the charade.

They all dutifully troop down to the shore. And wait. And wait. And this is where the problems surely begin. For the tide rises—inexorably, yes, but slowly too. Really slowly. Is water up to the ankles sufficient to convince the courtiers that their king is subservient to God as he wishes to show? Or must he be waist deep? How far does the water have to rise for the king's case to be proved? How long does this take? I have stood on the beach at the spot in Southampton where

---

Our Island Story provided much of the raw material for W. C. Sellar and R. J. Yeatman's priceless spoof of British history, *1066 and All That*. In their version, "Canute, an Experimental King," is informed by his courtiers that "(owing to a misunderstanding of the Rule Britannia) . . . the King of England was entitled to sit on the sea without getting wet. But finding that they were wrong he gave up this policy and decided to take his own advice in future."

King Canute is reputed to have tried to hold back the waves. My throne was a folding picnic chair set up to face the water a few feet in front of me. The tide rose just as it did a millennium ago and was soon lapping at my feet, then slowly creeping up my shins. It took about thirty minutes to get this far. What happens during this hiatus? Do Canute's courtiers stand patiently by, confident that their king's divinity will show itself now that he has embarked upon this proof? Or do they argue, with him, or amongst themselves, as the water begins to soak the hems of their cloaks? I imagine the scene somewhat along the lines of Leonardo da Vinci's *Last Supper*, with animated dispute going on all around the calm central figure. But there is no noteworthy painting of the scene. Engravings made for the history books tend to show the whole party firmly on dry land, with Canute merely holding his hand out to the sea like a traffic cop.

In its own time, the tide does as the tide must. "I am your lord and master, and I command you not to flow over my land," the king intones. He, at least, knows it is an act. How does he speak these words? Does he humor his court and play it deadpan? Or does he say the words with a wry smile, with an ironical raising of the eyebrow, even a sardonic curl of the lip? "Go back, and do not dare to wet my feet." The tide naturally ignores him, and we are left to presume that his entourage return to their offices duly chastened at having witnessed the scenario.

The story continues to supply one of the most popular examples of tide as a metaphor. Journalists today refer to politicians or officials as "acting like Canute" if they are seen as trying to stop an irresistible force. In doing this, of course, the reporters willfully mistake Canute's true intent, which was to fail in this operation. But sometimes a metaphor is more useful than a fact, even to a journalist. Canute has appeared in recent headlines describing everything from the position of the commercial music industry in the digital age to the stance taken by some priests against gay marriage. An especially evocative usage, in which the literal almost overtakes the metaphoric, was seen in June 2012 when coastal businesses in North Carolina lobbied—successfully, it is alarming to report—to have any reference to rising global sea levels expunged from state legislation. "NC Bans Sea Level Rise; King Canute Unavailable for Comment," noted one blog drily.

The stunt is supposed to have occurred in 1028, although the location is less certain even than the site of Julius Caesar's landing in Kent, and on this occasion, too, there is no contemporary account to refer to. A plaque on the wall of the former Canute Castle Hotel on Canute Road in Southampton records:

NEAR THIS SPOT A D 1028
Canute
REPROVED HIS COURTIERS

although it may have been Bosham in Sussex, or even Thorney, an island in the River Thames at Westminster long since lost to the tides, where the event took place.

The original source of the story is *Historia anglorum* ("The History of the English People"), the chronicle of Henry, Archdeacon of Huntingdon, who was writing about a century later than Canute's reign (1016–35). Henry drew together information from various Anglo-Saxon chronicles for his account, but he is likely to have taken the Canute story from poems in Old English. He writes that Canute "gave orders for his throne to be placed on the seashore. . . . And he said to the rising tide . . . 'You are within my jurisdiction, and the land on which I sit is mine; no one has ever resisted my command with impunity. I therefore command you not to rise over my land, and not to presume to wet the clothes or limbs of your lord.'" This is clearly the source used by H. E. Marshall.

The king's purpose had been to demonstrate not his omnipotence, but his powerlessness before God. After the exercise was completed, he is supposed to have explained, "The power of kings is empty and superficial, and . . . no one is worthy of the name of king except for

Him whose will is obeyed by Heaven, earth and sea in accordance with eternal laws." Henry of Huntingdon adds that "he took off his golden crown and never put it on his head again."

Such eccentric behavior by a national leader might be thought to be tempting fate, unless he was extremely confident of the loyalty of those around him and of political stability in the land. What was he really playing at?

Cnut (to give him the spelling preferred by today's scholars) was the first and only Danish king of England. His invasion force arrived on the Kent shore in 1015, and then moved swiftly by sea around the south coast to subdue Wessex and take control of the country. Cnut consolidated his position, appointing earls who were shrewdly split between native Anglo-Saxons and Scandinavians. Astute as a politician, he brought stability to a land that had suffered upheaval and corruption under the earlier king, Aethelred. Historians today take the remarkable absence of documents detailing the kind of discontents of which history is typically composed as evidence of his comparatively untroubled reign.

It is also significant that Cnut was a Christian, whereas his father, Sweyn, had been ambivalent about the new religion. Every Nordic leader at this time employed *skalds*, poets whose job it was to devise memorable verses relating his notable exploits—the predecessors of today's spin doctors. Their refrains, repeated and passed among the king's subjects, could be relied upon to build a favorable profile much like a modern brand slogan. Cnut's skalds laid emphasis not so much on the king's might in battle as on his beneficence, drawing an explicit parallel between his pacific and protective role over his kingdom and God's role in the kingdom of heaven. An important subsidiary effect of this message was to insinuate the implicit comparison between the king and God into the general social discourse. In this, Cnut was "a master of public relations," according to Roberta Frank, a scholar of the period at Yale University: "The parallelism of God and prince in these refrains seems to have been a kind of Cnutian leitmotiv."

However, it seems that the constant reiteration of "Cnut" and "God" in the same verses may have done its job too well, encouraging in the king's more credulous and sycophantic followers an inappropriate belief that he truly possessed divine powers, and this perhaps is what eventually forced him to stage the seaside demonstration of his limitations. It was for this reason, according to Frank, that the memorable but generic story of a king stemming the tide attached itself to Cnut as opposed to any other leader.

And what of the tide in all this? It is surely significant that the king chose the tide for his demonstration rather than pretending to tame a flowing river, or a wild beast, or some aspect of the weather. The tide is different from these things in that it appears to obey at least some rules. This was understood not only in the experience of coast dwellers, but also in an academic sense. In 725, the Northumbrian monk and historian Bede had discussed the tides in his book *On the Reckoning of Time*, noting the coincidence of the highest tides with the new moon and full moon, though with no understanding of any mechanism linking the two. This insight helped to dismiss earlier ideas that the tides arose by the action of unseen geysers. The tide was a force of nature, then, but one with a rhyme and reason dimly discernible to humans. For early Christians, the tide might seem the creation of a reasonable Christian God in comparison to the rivers and animals, which frequently retained the names of the old Norse gods.

Cnut is surely also thinking politically. In looking seaward, he takes the opportunity to remind his court, and through them his English subjects, of his jurisdiction not only over English lands but over a vast kingdom overseas, which by 1028 included not only Denmark, but also Norway and part of Sweden. He may not be the master of the tides, but he is the ruler of the most extensive kingdom in Europe.

A TIDAL VOCABULARY

You might expect our words for tidal phenomena to have come, like many place names in the east of England, on the waves of Nordic invasion. In fact, they came mostly with the earlier influx of the Anglo-Saxons. One of the busiest Anglo-Saxon landing sites was at Ebbsfleet in Kent, a place whose name seems to refer to the fast-running tide.

Bede wrote in medieval Latin, using terms such as *malina* or *malinae* and *ledon* or *ledones*, which express the idea of "lively" and "dead" waters, to mean what we now call the spring and neap tides, respectively. Other words were used to refer to phases in the tidal cycle to which we no longer give a name, such as *dodrans*, a word indicating a proportion of nine-twelfths and used to refer to the time when the tide begins to flood. In Romance languages, a neap tide is often still a "dead sea."

But in English, any Roman influence was swept away along with the preexisting Celtic tidal vocabulary, by the Anglo-Saxons. The word *tíd* could express the concept of the sea's rhythmic motion but was synonymous primarily with times of significance. The tide was generally spoken about not in this conceptual way, but in terms of the *flód* (flood) and *ebba* (ebb), which derive from Old Frisian, the early language that ultimately gave most of northern Europe its tidal terminology. Compounds of these words were used to express tides of a particular kind: *apflód* being low tide; *héahflód*, high tide; *népflód*, neap tide; *fylleflód*, spring tide.

The word "neap" does not arise on its own, as opposed to in compounds such as *népflód* and the later *níptíd*, until medieval times, except for one tantalizing occurrence in an Old English document known as the Caedmon manuscript. The text, thought to date from about 1000, is a retelling of the book of Exodus in the manner of a heroic epic such as *Beowulf*. It relies for much of its impact on descriptions of a vividly mobile sea that surely relocates the Israelites' escape from Egypt as something that might have happened on the English coast rather than along the edge of the Red Sea. The original lines, *Mægen wæs on cwealme fæste gefeterod, forðganges nep, searwum asæled*, might be translated as follows: "The force [Pharaoh's army] was bound for death, their feeble (*nep*) progress hampered by arms." But it is quite possible that *nep* here is also a poetic contraction used to refer to the state of the water, as well as the Egyptians. (The Bible nowhere makes any explicit mention of the tide, which is not all that surprising, given the small tides of the seas in the region of the Holy Land. The relevant passage in this case runs, "And the Egyptians pursued, and went in after them to the midst of the sea.")

The adoption of the word "spring" (taken from its still-used other meaning of an upwelling of freshwater) for the highest tides was a medieval innovation, which has confused those ignorant of the tides ever since into thinking that higher tides should be expected in that season of the year.

The ultimate meaning of *flód* is clear enough, since it is still consonant with flood in its general sense of "too much water." Its converse, *ebba*, derives ultimately from the Germanic prefix *ab-*, meaning "off or back." Etymologists note, however, that it is impossible to trace more technical tidal terms, such as "neap" and "spring," further back in time—presumably because these nuances of the tide had not yet been observed and so did not need to be named.

The route by which these words spread is surprising. Many tidal concepts are simply absent from Old Norse. We are used to the idea that the English language has been enriched by Old Danish and Norwegian. But on this occasion, it seems that the Scandinavian vocabulary was largely acquired from Anglo-Saxon England, where the tides are so much greater, after the Viking invasions.

BRIHTNOTH'S LAST STAND

MALDON, ESSEX

	LW	HW	LW	HW
August 10, 991		10:23		22:48
		2.7 m		2.9 m

Knowledge of the tides was clearly an important strategic asset during the waves of incursion that occurred along Britain's shores in the early Middle Ages. Chronicles tell of many military engagements on the coast. The Vikings began making raids on Britain in 793 and, finding the country poorly defended, they continued the strategy for some time, before eventual invasion and colonization.

In 896, King Alfred the Great is said to have defeated the Danes in a sea battle in the English Channel. Alfred used nine new vessels, quite possibly of his own design, which were twice the size of the Danish ships, as well as swifter and steadier, to blockade six Danish raiding ships in harbor. The confrontation occurred somewhere in the area of the Isle of Wight, according to the *Anglo-Saxon Chronicle*. Few places along the coast here correspond with the topography described in the *Chronicle*, but the writer and yachtsman Hilaire Belloc identifies the notorious Bramble Bank in the Solent at the mouth of Southampton Water, which sometimes dries at low tide, as one place where the set-to may have happened.

Three of the Danish fleet sailed out to confront the English on the falling tide. Two were quickly captured and their crews slaughtered, while the third ship managed to flee. In the confusion, however, three of the English ships ran themselves aground close to where the remaining three Danish vessels were pulled up on the beach, separated from them only by a small inlet where the water was still ebbing away. Worse still, the remainder of the English fleet ran aground on the far side of the seaway and, with the tide still falling, had no hope of coming to the aid of their compatriots.

As the tide fell further, the Danes rushed from their boats, fording the inlet on foot to engage the three stranded English crews on land. However, the English had the best of it, helped out by Frisian mercenaries among their number. When the flood tide came, the Danish ships refloated first and made their escape, hoping to flee around the coast to Viking-held East Anglia. In the end, two of the ships were captured on the Sussex coast, and only one made good its escape.

Among his many accolades, Alfred was later styled as the founder of the British navy. Nevertheless, although they had secured a victory on this occasion, his captains did not in general distinguish themselves in seamanship. Even from the favorable reporting in the chronicles, it is clear that the English seamen had great difficulty in handling their new vessels in demanding circumstances, and that they had conspicuously failed to use local knowledge of the tides to their advantage.

A century later, matters had hardly improved for the Anglo-Saxons. In the summer of 991, a large Danish fleet made repeated raids along the coasts of Kent, Essex, and Suffolk. On August 10, Danish longships sailed up the River Blackwater to Maldon.

This time, the lay of the land and sea is quite clear. A gray estuary slithers in from the far North Sea. In front of me, rising out of the river mud, is a low island, no more than half a mile in breadth. Land and sea are negotiable here: the mainland banks of the river and the island itself are infiltrated by meandering creeks and sluices—so much so that at high tide the island must become many islands. But there are a few firm fields for pasture in the center, and it is a mapped feature of the landscape, called Northey Island—"the Wuthering Heights of Essex," according to some wag producing copy for the National Trust website. On the southern shore, a farm track leads to a causeway that crosses to the island only at low tide. By the side of the track, a sign tells me this is the earliest "Registered Battlefield in England."

The Danes sailed up the river on the flood tide and beached their

boats on this island. Separated by swift tidal channels from the mainland to the north and south, it was a strategic location. The Vikings' plan must have relied on local knowledge from previous raids, as well as knowledge of the tides, in order to time their arrival at the island for maximum advantage. Their fleet would be safe from attack at least while the tide was in. The Danish troops would have a few hours to disembark and regroup for the inevitable battle ahead. Meanwhile, on the shore, King Aethelred's local sheriff, Ealdorman Brihtnoth, amassed his men ready to repel the raiders.

The Battle of Maldon is recounted in detail in surviving fragments of a contemporary poem, which makes it clear that the movements of the tide were once again a critical factor in the outcome. The Vikings shout from the island that the English must pay tribute or face the consequences. Brihtnoth answers the "sea-robbers" robustly: "For tribute they're ready to give you their spears." But then the tide intervenes:

> Nor could for the water, the army come at the other,
> For there came flowing, flood after ebb;

Locked were the ocean-streams, and too long it seemed
Until they together might carry their spears.

The standoff must have lasted for several hours, until the tide receded again. Who used the time better? The Vikings, waiting with their boats safely pulled up on the island, or Brihtnoth's men on the shore, with ample opportunity to call up reserves and supplies from further inland? It begins well enough for the defenders. When the sea withdraws, the first few Viking warriors begin to cross the causeway from the island and are felled by the English as they advance. But then Brihtnoth "in his overweening heart" summons them all across the causeway to do battle fair and square:

Over the shining water they carried their shields;
Seamen to the shore, their bucklers [small shields] they shouldered.

In the end, Brihtnoth's army proved no match for the invaders. His men fought valiantly, but when Brihtnoth himself was mortally wounded, they realized the game was up, and many of them scattered or fled. The humiliation was to have severe consequences for the whole of the large part of England controlled by Scandinavia, marking the beginning of the systematic imposition of the hefty land tax known as Danegeld. On this occasion a levy was demanded and paid of 10,000 Roman pounds in silver.

Yet it would seem that Brihtnoth could have held the Vikings off almost indefinitely on the island under siege conditions, picking them off one by one if they dared to cross the causeway. Why did he not do this? Perhaps he feared that they would sail further upriver on the next tide to attack somewhere else where there were no ready defenses, and hence saw the urgency of drawing them into battle here. Or perhaps it was just bravado. Academics squabble about the exact meaning of the Anglo-Saxon *ofermod*, translated as "overweening heart" above, which might indicate the positive virtue of great spirit or a less positive quality of sinful pride. Either way, on that long-ago August day, Brihtnoth established himself as a progenitor of that favorite English type, the heroic failure, fully worthy of celebration in tapestry and poetry at the time, and indeed still so today, as the subject of noisy ballads by various black-metal bands.

On the battle site today, wooden stumps stand here and there in the marshes, uneven in height, like memorials to the fallen. The water still shines on the causeway. A little distance away along the riverfront in Maldon town is a fine bronze statue of Brihtnoth standing with his sword raised bravely against the sea.

THE BATTLE BEFORE

The story of the English king Harold's defeat to the Norman William the Conqueror at the Battle of Hastings in 1066 is well known. Keen students of history will remember also the battle a few days before that, when Harold defeated the Norwegian pretender to the throne, Harald Hardrada, at Stamford Bridge in Yorkshire, nearly 300 miles away to the north. What is less well known is the story of the battle before *that*. Yet, the Battle of Fulford, not far from Stamford Bridge, may in the end have had more influence on the development of England than its famous successors, and here, too, a decisive role was played by the tide.

Fulford is where Hardrada's invasion force first ran up against the English defenses. His fleet sailed up the Humber and the Ouse Rivers toward York, the northern capital, mooring up at Riccall, where the troops disembarked and continued on foot. They met a combined detachment of Mercian and Northumbrian soldiers who might usefully have drawn the invaders on to well-protected York, but instead chose to take on the Norsemen on boggy ground on the banks of the Ouse. The invaders mustered along one edge of a creek running off the main river, known as Germany Beck, the English along the other, both waiting for the tide to fall. Local historians have computed the likely tide times and heights for the day in question and have concluded that here, as at Maldon, the state of the tide probably delayed the conflict. The more experienced Norwegian fighters managed to cross the water first, outflanking the English and driving them into the muddy creek, where many were slaughtered. (More recently, the area was a battleground again as local campaigners fought unsuccessfully against plans to build hundreds of new homes on the historic site.)

Having triumphed here, Hardrada's army was then taken by surprise by the arrival of King Harold's army marching up from the south of the country just a week later. The Battle of Stamford Bridge marked the end of Norse domination in Britain. The irony is that Harold and his depleted and exhausted army then had to race south again to meet their better-known date with destiny at Hastings. If the Mercian and Northumbrian armies had prevailed at Fulford, Hardrada's men would have been greatly weakened, and King Harold might have scored an easier victory over them, leaving his army in a better state to take on the Norman invaders from France; English history might have been very different.

It is odd that the English, living on some of the most tidally active shores anywhere, should so often find the tides put to use against them by enemies from across the sea. The tides in Norway and Denmark are considerably less than they are along most of Britain's coast. Yet, in an age when boats had to be rowed if there was no wind for sailing and could attain speeds of only a few knots—comparable with typical rates of tidal flow around British shores, particularly those found in the estuaries chosen for the assaults—it is clear that the superior seamanship of the Norsemen outweighed the importance of any local tidal knowledge the English might have possessed.

Even with faster vessels and more sophisticated military strategy, the tide continues to be a factor that cannot be ignored in warfare, playing an important role in amphibious assaults even today. Indeed, for as long as combatants choose to wage war by land and sea, the tide is likely to be judged afterward as having allied with one side or the other.

BEDE'S SENSE OF TIME

These defeats are the more surprising since, by the time of the Viking invasions, the English were indeed in possession of superior knowledge of the tides, at least in a theoretical sense.

One of the very first Viking raids had taken place on St. Paul's monastery at Jarrow on the River Tyne in 794. It was here, more than a century earlier, that a seven-year-old boy had been left at the door, who grew up to become a Benedictine monk called Bede.

Bede proved to be a prolific polymath, writing works of history, theology, grammar, natural history, and hagiography. He also had a decidedly scientific curiosity. When he saw the sunlight streaming through the stained-glass windows of his abbey, his mind wandered into speculations on optics, for example. The titles of his early works bear this out: *On the Nature of Things* and *On Time*, both dating from around 703. In fact, both of these treatises demonstrate Bede's preoccupation with the concept of time, the first of them even being organized by chapters based on units of time; his discussion of tidal lore here comes in the chapter concerning the month. In 725, Bede followed *On Time* with a more detailed work, *On the Reckoning of Time*, which has rather more to say about the tides. It was this work's promotion of using the date of Christ's birth to calculate the year, in place of the confusing system of using the regnal year of the local monarch, that gave us the anno Domini system we still use today. Finally, Bede's most famous work, *The Ecclesiastical History of the English People*, of 731, also betrays his scientific mien and his preoccupation with the shores of his homeland. In fact, it opens with estimations of the length, breadth, and circumference of "Britain, an island of the Ocean, which formerly had the name of Albion."

Before there were clocks, the idea of measuring time was nearly meaningless. People measured the hour in an approximate way from the height of the sun in the sky, or from other related natural occurrences, such as birdsong or breezes. For fishermen and other coast dwellers, matters were a little different; for them, quite simply, the tides were the time, and their days were planned according to their ebb and flow, with the pattern of one day simply repeating itself the next day a little later, until the day a fortnight or so later when the whole cycle was repeated.

The science of time reckoning was more important in monastic communities because it provided a way of fixing the dates of Christian feast days. Bede never went on pilgrimage or on travels involving sea voyages, so why he should be curious about the tides seems at first a mystery, but this interest also follows naturally from his theological problem solving. In 703, there was no agreed-upon formula for calculating the date of Easter. This most important date in the Christian calendar is based in part on the phase of the moon, but this fundamental astronomical datum must also be reconciled with the invented calendars of humans. A complicated procedure called the computus was used for this purpose, merging elements of the thirteen-month Jewish lunar calendar, in which Passover falls on the day of the first full moon of the first month, and the familiar twelve-month Roman calendar.

Bede's interest in the sun and moon arose initially from this need to address the Easter problem. Stationed as he was on the frontier between Roman and Celtic Christianity, he was well placed to offer a universal solution to the problem if he could. He was also familiar with the great works of classical antiquity in the well-stocked library of his monastery, where he was employed in preparing abridgments of key manuscripts for the dissemination of knowledge to less fortunate institutions. He was also in contact with Irish monks, who had their own ideas about the date of Easter, as well as their own observations of large tides.

On the Reckoning of Time, written at the request of Bede's monastic students, extended and developed the ideas set forth in *On Time*—as well as rebutting an accusation of heresy that had been made against the earlier work. (In the course of his calculations, Bede had come up with an unorthodox estimation of the age of the earth, which had upset some ecclesiastical authorities.) Here, Bede integrated the traditional computus with a clearer understanding of the astronomical and cosmological context on which it was dependent. He showed, for example, how the Jewish and Roman calendars converge every nineteen years, coincident with the slowest rhythm observable in the tides. He also included his findings concerning the moon and its effects made by various direct and indirect methods, noting "this union of the ocean with the orbit of the moon," and collating folklore about the influence of the moon on the size of oysters and the best times to sow and reap.

Bede even provided a metaphorical glimpse of the attractive force of gravity: It is as if the sea is "unwittingly drawn up by some breathings of the moon," he wrote. "When the moon's force has ceased," it is as if the sea "was poured back into its own proper basin." More significant, he noted that the time of high tide falls back a little each day, in time with the hour at which the moon rises. He measured this delay to be 47½ minutes. And he put on record the relation there seemed to be between the heights of tides and the phases of the moon.

Bede must have measured the tides himself on the Northumbrian coast where his monastery was situated. But his real breakthrough was to solicit similar data from other locations up and down the British coast. He was perhaps inspired to do this by comparing anecdotes with visiting monks from other seaside monasteries, such as from the remote island of Iona on the west coast of Scotland. In other words, he used the existing monastic network to make a scientific survey. It was, according to one modern Bede biographer, "what might plausibly be called 'research.'"

The data Bede gathered showed clearly for the first time that the tide might be high in one place and low in another at the same instant. "For we know, we who live in different parts along the coast of the British Sea, that where in one place the sea begins to flow, at the same time in another place it begins to ebb." Along Bede's own stretch of coast, his studies revealed in particular that the tide came earlier in the north and later in the south. The moon, of course, will rise and set at about the same time in all these places because they lie close to a north–south line. It might seem, therefore, that the tide can have no correlation with the appearance of the moon. But the correlation is still present. It is simply that, in each place, high tide may be observed to coincide with the moon being at a certain angle in the sky, but the angle happens to be different. The moon is not necessarily high in the sky at high water, or low on the horizon at low water, even though its track across the sky is apparently synchronized with tidal movements.

Bede's findings refuted those who still clung to the classical notion that the tides were the expression of some kind of life fluid of the earth that expanded and contracted everywhere uniformly, or that they were produced by a coordinated flow of waters in and out of the mouths of the world's rivers. His science traveled by word of mouth, albeit sometimes in garbled versions, but was not published until the sixteenth century.

Bede came to be known as "Venerable" in the century after his death, and long afterward, in 1899, he was made a saint. He has the possibly even greater distinction of being the only Englishman to be found in the "Paradise" of Dante's *Divine Comedy*. Although he had made what we would now call a scientific discovery, the context of his work was entirely devotional. He saw the wash of the tides not just as a natural phenomenon, but as a purifying influence. Science and religion are not so easily separated at this time, and it is easy to believe that the manifest power of his inductive reasoning must have contributed to the high esteem in which he was held as much as his religious learning did. As the oceanographer David Pugh observes, "An ability to predict future events, particularly those of practical importance, must inevitably have attracted some veneration to those who practised the art."

KNOWABLE MYSTERY

A century after Bede, Abu Ma'shar al-Balkhi, a Persian-born astrologer working in the Abbasid court at Baghdad, made still greater strides in the theory of tides. Albumasar, as he was known to Western scholarship, wrote a major astrological treatise that became available to Europeans through a Latin translation in the twelfth century. It is possible that we owe our recovery of Aristotle's science to Abu Ma'shar, whose Arabic versions of his works were also translated into Latin around this time. Abu Ma'shar stood with classical theory in identifying the moon as the chief agent of the tides, postulating that its light heated the water and made it expand.

As with Bede, Abu Ma'shar's inspiration was religious as much as scientific. Light to him was not only a physical property of the sun and moon, but a sign of the divine and the source of the human soul. The power of the light of the moon over the oceans was a demonstration of Allah's omnipotence. Abu Ma'shar also noted a number of new relationships, such as the additional contribution to the height of the tide made by the sun on days near the full and new moons. He observed the coincidence of the highest tides with times when the moon appeared large in the sky because it was at its closest to the earth. He also recognized the importance of the angle of the moon above the horizon (its declination) in governing the tides. In all, he identified eight distinct causes of the tides, most of which were correct in principle and form an integral part of tidal calculations today.

Already, in these early medieval times, there was an emerging sense that the tides are a complex phenomenon, subject to many influences, each of which must be understood in order to be able to explain and predict them properly. Abu Ma'shar's theory of heating by moonlight, for example, did not explain why there was also a high tide each day when the moon was below the horizon.

Abu Ma'shar implicitly recognized the difficulty of the problem by counting the winds and various aspects of coastal topography as influences on actual tide heights and times, although these are not strictly components of astronomical tides. The sometimes large variation in the times of the tide even between places not very far from each other along the same coast or lying across quite narrow bodies of water, such as the Irish Sea, from where Bede obtained some of his measurements, appeared to suggest that there was no simple or universal rule for predicting the tides. But it was equally obvious from local experience that the tide could nevertheless be predicted accurately enough to be of great practical use to seafarers without a highfalutin theory. If, as a medieval merchant or fisherman or warrior, you can predict the tide, then there is really no need to inquire further into its basic nature. That job would have to await the coming of new scientific minds some centuries in the future.

Naturally, the most dramatic and most regular features of the tide were the ones for which thorough predictions were first made, especially when they happened to occur in populous places and commercial centers. For example, the Chinese have long celebrated the tidal bore that surges up the Qiantang River toward Hangzhou. Known as the "Black Dragon," it is the world's largest river bore. Observing it has been a spectator sport for thousands of years, and today there are even pavilions erected as permanent vantage points at strategic locations. The biggest celebrations take place at a festival in September, when the highest tides often come. Other monuments have been built along the river in order to propitiate the dragon, which has claimed many lives down the centuries. The attraction and danger of the phenomenon led the Chinese to produce the first known predictive tide table in 1056, giving the time of arrival of the bore each day and its likely height.

The predictions are not always accurate even today. In August 2013, the bore wave greatly exceeded the height predicted, owing to the additional effect of a recent typhoon, and overtopped barriers where spectators had gathered to watch, causing them to flee in alarm. In August 2014, a "supermoon"—which occurs when the lunar orbit is at its closest approach to the earth—contributed to another exceptionally high tide, with massive waves of filthy river water again crashing over the concrete embankments, soaking thousands of sightseers and injuring many.

The first table for predicting the times of ordinary tides was made around 1220 for the tides at London Bridge (a location still used in tide tables today) by one John of Wallingford, a Benedictine monk at St. Albans Abbey. Inked in heavy italics, it looks remarkably like a modern table, with daily times of the "fflod at london brigge" given in hours and minutes in one column, and another column giving the times at which the moon passes through north in the sky, which happens three hours before each high tide. The times of both events advance by precisely forty-eight minutes each day. Unlike today's tables, however, John of Wallingford's table gives only one tide time per day, each day's other high tide being near enough to twelve hours later, and therefore having the same twelve-hour-clock time, to need no spelling out. With its clear relation to both clock time and the position of the moon, such a table might be used by city traders and mariners alike.

THIS, THEN, was the state of understanding of the tides around the world before the advent of modern science. Subject to divine regulation, as Cnut was happy to concede, the tides were knowable to a certain degree—often to a useful degree—in their times and heights. But they were also liable to unpredictable excesses and clearly contained many mysteries. It would be some centuries yet before those mysteries truly began to be solved.

4

NO PATH
THROUGH WATER

ANOTHER PLACE

I am standing on Crosby Beach near Liverpool staring out to sea. I am not alone. With me are a hundred cast-iron men doing the same, spaced at intervals up and down the slope of the beach over a distance of a couple of miles, all implacably surveying the outer approaches to the River Mersey.

Unlike them, I am thinking about the walk I have arranged to do the next day, some miles further along the coast, across the tidal flats of another estuary, at the north end of Morecambe Bay. The sands there are exposed only at low tide and are full of dangers.

The iron men are the work of the sculptor Antony Gormley. He has called it *Another Place*. The tide is just beginning to rise when I arrive, and the first row of figures is already thigh deep in the water. It is a cold day with a stiff onshore breeze, and their "paddling" has an air of unreality about it. I set off along the beach to inspect one of the men close-up. It is farther than I thought—the familiar trick of scale in the sea light. I look still farther off. Are the distant figures more Gormleys or real people? It is hard to tell.

The sea has reshaped the beach in the few years that the men have stood here so that some of them now stand on little plinths while others are buried in sand up to their knees. The dunes at the top of the beach, I notice, are anchored in place with the help of last year's Christmas trees. It is not only the sea that is in constant motion here; the land is also on the march, piling and repiling itself ever further inland.

Some of the iron men have been defaced with paint. One of the rusty torsos has been given a bikini; another wears a condom. What would the artist have us believe they are thinking? Where are they looking with their unblinking seaward gaze? What is the "another place"? A land over the sea most obviously, in this city of immigration and emigration. Or perhaps an undersea world, a marine topography to parallel our terrestrial one. Or the sea itself as a place. Or is the artist driving at something more spiritual? The beach is constantly remade as another place too. The men need only stay where they are to be somewhere else.

My contemplative mood is shattered by the noise of bleeping diggers rearranging the sand in readiness for the summer season. The world moves on. A wind farm offshore ranges an army of Wellsian white monsters against Gormley's rusting human figures. Perhaps "another place" is the past.

Two lifeguards have driven out onto the beach to keep watch as the tide comes in. They sit with their windows up in their truck. I ask them whether they have much to do. They explain that the main hazard is the sandbanks, which people don't notice are there at low tide, and which can leave them cut off from the beach when the tide rises. At first, the sculptures were an additional nuisance for the lifeguards, as people would report swimmers in distress. But now, it seems, everybody is used to them. The local council voted to have the artwork removed after a while, citing health and safety dangers. But public pressure means the iron figures are now here to stay.

I watch as the water rises. Then I spot one of the figures immersed up to its neck. It is the size that forces you to focus for a moment. Is it a seal, or a buoy marking the position of a lobster pot on the seabed?

Or is it a real person swimming? Or drowning? Then I lose it amid the dark shadows of the waves, and it goes under. I imagine myself caught out in Morecambe Bay.

COAST LINES

The coast of the main island of Great Britain, the coast I am presently standing on, is nearly 18,000 kilometers in length. In fact, it's 17,819.88 kilometers, according to Ordnance Survey, the national mapping agency. The Venerable Bede thirteen hundred years before estimated that "its whole circuit amounts to four dozen times 75 miles," which would be a mere 3,600 miles all told. In truth, though, the length depends on the tools used to measure it. Ordnance Survey's measurement is based on its maps drawn at a scale of 1:10,000. The same distance measured off a less detailed map comes out at rather less because many of the crinkles in the coastline—muddy inlets and rocky promontories—no longer appear. They are swallowed by the thickness of an ink line. A walker striding along the waterline (surely this is the best measure to take in a book about tides) might count so many million paces, whereas an assessment made, let us say, from a satellite map with a resolution of 10 meters would come out at rather less, because the weaving route of the walker across each of those 10-meter segments would be lost, approximated as a straight line instead. I have seen a figure of 150,000 kilometers used for the perimeter length of all the continents in global calculations of the dissipation of tidal energy, but this must be an even grosser approximation; irregular as its outline is, Britain surely cannot claim more than 10 percent of all the world's coastline.

The Polish-born pioneer of fractal geometry Benoit Mandelbrot chose the British coast to illustrate the phenomenon he called the "coastline paradox" in a famous paper of 1967, giving a name to the property of certain things that the more closely you observe them, the more intricate they seem. He later coined the word "fractal" to characterize geometries of irregularity and roughness that conceal unexpected order because their constituent shapes—a blobby eruption here, a jagged fissure there—in fact recur at different scales. Fractal geometry provided a formal way to describe many unruly phenomena, from the shapes of clouds to the jitters of the stock market.

A tidal coast multiplies the difficulties of knowing its length. Many lines can be drawn between land and sea. Now it is not only a problem of how closely you look—whether you take into account each rock or each grain of sand that the water laps around—but also the line is constantly moving. Ordnance Survey indicates the high-water mark as a black line. The low-water mark is not represented by a line but is simply where the tone of ink used for the shore (a pale ocher for sand, greenish for mud) gives way to the blank wash of bathroom blue that somehow manages to indicate both the great extent of the sea and our great ignorance of it. The terrestrial mapmakers do not specify which of the many available high- and low-water datums they are using—average highs and lows, or average spring tides, or the greater extremes reached by what are known as astronomical tides. However, that doesn't matter much. It seems obvious that the low-tide perimeter of an island will be longer than the high-tide one simply because it encloses more land. But, in fact, this is not necessarily so.

In terms of fractal geometry, the low-tide line may be essentially similar to the high-tide line, or it may not. A straight stretch of uniformly sandy beach will have the same length at both states of the tide, whatever the level of detail one chooses. The water lapping on the sand grains makes a similarly complicated line in each case. A rocky coast may have the same fractal quality at high and low water too, if the rocks happen to provide the same little inlets and craglets and pools at both extremes. But this self-similarity of the coastline at different states of the tide cannot be relied upon. A high water that laps the edges of reticulated salt marshes and penetrates far up into meandering tidal estuaries to create a very long coastline may recede far out across smooth mud or sand to leave a low-tide line that is straighter and much shorter.

Fortunately, these crinkles and creases are not a factor when it comes to measuring something of more than theoretical interest: the area of the intertidal zone that lies between these two lines. This is what matters to the seaweeds and mollusks that make this their home. The intertidal zone is one of the largest and richest of habitats. But it is also, unlike woodland or savanna, uniquely distended, stretched, and squeezed, and it is exceptionally fragile and vulnerable to human thoughtlessness.

What is the extent of the intertidal zone? The length of the coastline is clearly the major dimension, although thankfully the calculation is simplified because we are now measuring the length not of an infinitely thin straggly line, but of a broader band that allows us to lose the unnecessary detail at its edges. The dimension nominally perpendicular to this line—that is, the shortest distance between the high- and low-water marks at any point along the coast—is the more troublesome variable now. This dimension varies according to the local geology and also the local tidal range. A sea cliff in an area of little tide will have a roughly vertical intertidal zone spanning a few meters at most, and perhaps only a few centimeters somewhere like the Mediterranean, whereas a gently shelving beach in a place where the tides are greater may stretch for miles between the high- and low-tide lines. The extent, orientation, and composition of this zone, and the way that it is traversed by tides and splashed by waves, are crucial to the kind of life it can support.

If we cannot calculate the length of the coast, can we at least make a sensible estimate of the size of the intertidal zone? It seems this has not been done satisfactorily either. There are too many factors that won't stay still: the curve of the coast, the slope of the beach, the range of the tide. But I would not be at all surprised to find that, if the coasts of all the countries of the world were unraveled, and their tidal ranges factored in along their length, the United Kingdom would come out on top. Countries such as Norway have more indented coasts but experience smaller tides. Parts of Canada have greater tides, but other parts are hardly tidal at all. No, of all nations, I bet Britain has the largest intertidal zone in proportion to its land area. Do we notice? Do we care? Perhaps not. Perhaps that is just as well, and life here can go on relatively undisturbed in something close to a true natural habitat, perhaps the most natural one we have remaining to us, unless we enter the sea itself.

COCKLE SANDS

MORECAMBE, LANCASHIRE

	LW	HW	LW	HW
February 5, 2004	05:13	10:56	17:45	23:15
	2.1 m	8.6 m	1.8 m	8.6 m
February 6, 2004	05:51	11:29	18:22	23:50
	1.8 m	8.9 m	1.6 m	8.9 m

Morecambe Bay is the largest intertidal area in Britain, more than 600 square kilometers in extent, and containing the estuaries of five rivers. Its mud- and sand flats patched with seaweeds and eelgrass and crusted with invertebrates are the feeding grounds for shallow-water fish at high tide and huge flocks of wildfowl and waders at low tide.

It is important to humans too. People have always walked across this expanse at low tide because it is an irresistible shortcut. The route overland between Hest Bank near Morecambe and Kents Bank on the Cartmel Peninsula is nearly 40 kilometers, whereas the straight line across the sands is a mere 12, and promises to be easy going since there are clearly no hills in the way. Not all that long ago, farmers and artisans took their produce and wares to market this way. Crossings were made on foot and on horseback, in traps and carriages. Manor houses on the peninsula often employed carters to drive across the bay to buy produce in Lancaster. Occasionally, too, felons were escorted this way to the Lancaster jail.

This past is enshrined in law, which in turn is reflected in the modern Ordnance Survey map of the area. It shows the familiar dotted pink line of a public footpath running across the sands, a country walk that has taken an apparently mad detour and now heads straight as a bolt for miles across the uniform tint of the empty sands. On the map, it looks dangerously like a walk you might think of taking at any time. The route is legally a "byway open to all vehicular traffic" under the terms of the Wildlife and Countryside Act 1981, and so it is shown on the map. But you would be mad to take it in an ordinary car, which would soon be flooded or bogged down. Even on foot, the route is fraught with dangers, and it is strongly recommended that people not attempt it unless accompanied by a guide—a recommendation I have happily followed.

The day I have arranged to make the crossing dawns with a heavy drizzle. It is cold for May, and I am secretly hoping the walk might be called off. There is good reason to think it might be. It has rained heavily for most of the night, and this will add a large volume of freshwater to the ebbing tide, making the channels we will have to cross deeper and faster-running than usual. But as I drive down to Arnside, where the walking route starts, the cloud base rises and I see the empty sands spread before me. The prospect is almost inviting.

The bay is hazardous precisely because of these vast flats. It is not so much the large tidal range—up to 10 meters between low and high water—that poses the danger, but the fact that the tide rushes in unimpeded across these ironed acres, reaching speeds of 10 miles per hour or more, swiftly filling unseen channels and isolating low sandbanks. Subtle dips, invisible from the land, become watery traps. In some places, there also is quicksand.

It is almost impossible to believe that this huge bay can be filled so fast. The figures aren't much help. Using the crudest approximation, 600 square kilometers of land area multiplied by a 10-meter gain in height gives a total volume of incoming water of 6 cubic kilometers, or 6 billion cubic meters. The habitual comparisons with Olympic-size swimming pools are not even worth making. All this water must move in during just a few hours, and must then leave again just as promptly when the tide ebbs. To this must be added the freshwater joining the sea from the rivers. None of this flow is stable. With every tide, the fast-moving waters bring in new sediment and rearrange the mud and sand already lying on the banks. Over time, these things can produce major alterations of the topography. In the exceptionally wet spring of 1983, the major channel that for years had run close to Grange-over-Sands suddenly moved to follow the southern bank, where it runs still, more or less. For, I note that the channels I now see before me correspond in no way to those shown on my Ordnance Survey map, even though it has been revised in the last few years.

I KNOW IT IS AN ENVIRONMENT where it is best not to take chances because twenty-three cockle pickers were drowned here in 2004. On the afternoon of Thursday, February 5, a group of more than thirty young men and women who had come to Britain in search of work from Fujian Province in China drove far out onto the sands. As the short day turned to night, the tide began to come in. It was one day shy of full moon, and so the tides were springs. In haste, the pickers piled into the old Mercedes van they had used to ferry them to the cockle beds, but the vehicle was unable to pull free of the sand and was swiftly hemmed in by deep channels. As the freezing waters rose over the wheels and then up to the windows, the cocklers took refuge on the roof. Darkness had now fallen. One of the men raised the alarm using his mobile phone. Others made desperate calls to their families. For most, it was the last call they would make.

That night, several of the workers were rescued by helicopter, but the Royal National Lifeboat Institution (RNLI) hovercraft (the waters are too shallow and treacherous for a lifeboat), which arrived shortly afterward, found a sea of bodies. Twenty-one bodies were recovered over the next few days. One woman's skull was found six years after the tragedy. One man was never found.

It was an avoidable tragedy. The previous year a larger group of Chinese cockle pickers had been similarly caught out, but they were successfully rescued. A permit scheme, requiring those in charge of the picking gangs to take a course on safety in intertidal conditions, as well as to abide by rules restricting the harvesting of cockles in an effort to ensure sustainability, had been introduced just two months before. But the gangmasters paid little heed to these regulations, and confrontations with local fishermen forced the Chinese to search in more dangerous areas. In this case, evidence given by one of the survivors led to their Chinese gangmaster being sentenced for manslaughter and other offenses; he was jailed for fourteen years. In the wake of the disaster, further new legislation was introduced requiring gangmasters to be licensed, and there are now more patrols during the lucrative cockling season. In Morecambe Bay and in similar intertidal areas around Britain's coast, cockles are still a powerful lure—said to fetch £100 a sack—and there have been a number of large-scale rescues since 2004. It seems it is only a matter of time until the tragedy is repeated.

People living near the coast have always harvested the shellfish found in the intertidal zone; a vast sessile larder of high-protein food is not to be ignored. It is this same food source, unveiled and then miraculously replenished with each cycle of the tide, that also makes Morecambe Bay a vital habitat for hundreds of thousands of winter birds.

But it has always been a dangerous way of making a living, and marginal in more ways than one. J. M. W. Turner is one of many artists to have painted people journeying across the "Lancaster sands." His bleakly picturesque watercolors made nearly two hundred years ago contrast the bustle of people on the flat sands with the majestic Cumbrian peaks in the distance. The artist makes good use of the big sky to display his skill with atmospheric weather effects, but there is no sense of danger. Crossing the bay was then a routine hazard borne by many.

A more telling indication that gathering a harvest from the low-tide shore is always a hazardous activity is seen in Auguste Delacroix's painting, *Ramasseuses de coquillages surprises par la marée* ("Shellfishers Caught Out by the Tide"), of 1852, which hangs in a tiny island museum near Cherbourg in France. It is a mawkish composition, typical of the period, and Auguste patently lacks the exotic flair of his illustrious namesake and contemporary, Eugène. The canvas shows a group of girls trapped between the rising sea and a cliff hard behind them. The two oldest girls clasp their hands and pray for deliverance. There are two small children with them. One vainly tries to scale the cliff, while the other buries her head in the skirts of one of the older girls. Their baskets lie overturned, their harvest of shellfish scattered on the shrinking sands. Examining the detail, I notice that the older girls are wearing conspicuously large gold earrings, and I wonder if this is the artist's way of telling us that they are gypsies or the luckless and put-upon migrant workers of his day.

It is not only greed and exploitation that lead people into danger in Morecambe Bay. It is often just ignorance and foolishness. The mobile phone has undoubtedly saved many people stranded by the tide who would previously have had no way to alert the emergency services to their predicament, but in one case it left a testimony of almost unbearable pathos.

On January 5, 2002, a local man, Stewart Rushton, and his nine-year-old son, Adam, made an excursion onto the sands possibly to dig for bait for a planned sea-fishing trip. The tide was already on the rise, but they perhaps were not going to go far or stay out for long. On this occasion, though, the cold flood tide brought in with it a thick sea fog, and the two were suddenly unable to see their way to shore or even to see a safe way forward in front of them to slightly higher ground. The pair could only stay put as the tide rose around them. Using his mobile, the father called for rescue. Soon, the two—the boy now lifted onto his father's shoulders—heard rescue teams nearby, but despite screaming, they were unable to communicate their position in the fog. The boy reported, "My daddy's all right," as the water rose up to his neck. By the next call, seven minutes later, the phone was being held above the water; the rescuers could hear only the waves. The bodies could not be recovered until the next day, once the fog had lifted and when the tide was out. The coastguard refused to release the full transcript of the calls because it was too harrowing to be reported.

An inquest later called Rushton "reckless" and recorded a verdict of death by misadventure and, in the case of his son, accidental death.

CRUEL AND UNUSUAL PUNISHMENT

I am reminded that the transitory land between the tide lines has occasionally been chosen as a macabre execution ground. In the Solway Firth, a similarly expansive sandy estuary some miles north of here, in May 1685, during the "killing time" of Scotland's religious conflict, two women Covenanters—members of a sect of Scottish Presbyterians—were sentenced to death for refusing to swear allegiance to the official Episcopal Church. Men persecuted in this way were usually hanged. However, the method of execution chosen for Margaret Maclauchlan, a widow in her sixties, and eighteen-year-old Margaret Wilson was to tie them to stakes driven into the sand in the Bladnoch inlet, outside Wigtown, and then leave them to drown on the incoming tide.

Both women were offered the chance to renounce their faith. According to accounts of the execution, the older of the two was placed on a lower patch of sand, so that the water would reach her first, in order to give the younger woman more time in which to reconsider the error of her ways. But neither woman would say what was required of her. Margaret Wilson sang the twenty-fifth psalm as the waters rose, until she began to lose consciousness. Finally, one of the soldiers carrying out the sentence lifted her head so she could speak, and she was asked if she would pray for the king. She refused, and with that her head was pushed back under the water. The women were championed as the Wigtown Martyrs by Presbyterians, while Episcopalians queried whether the sentence was ever carried out. Nevertheless, a memorial erected to the "virgine martyre" Margaret Wilson in Victorian times gives the names of those who perpetrated the "cruel crime" against her and this verse: "Within the sea ty'd to a stake / She suffered for Christ Jesus sake."

Perhaps the executioners got their gruesome idea from Viking folklore. Stories in Norse sagas suggest that drowning on the rising tide was a standard method of judicial killing. A highly inauthentic—and surely the cheesiest—scenario occurs in the 1958 Hollywood epic, *The Vikings*, in which half-brothers Einar and Eric scrap for their father's throne in Northumbria. Showing off, Eric (Tony Curtis) lets fly his hawk, which blinds Einar (Kirk Douglas) in one eye. Eric is sentenced to death, but the omens are not favorable for a man to perform the execution. Instead, Eric is taken to a tide pool and lashed to a post to be drowned by the powers of nature. He is saved by an appeal to the god Odin, who causes the wind to abate the tide.

In *Peter Pan*, the pirate Captain Hook deposits the American "Indian" princess Tiger Lily on a rock in the hope that she will tell him the whereabouts of Peter, threatening that otherwise he will leave her to be covered by the tide. Of course, Peter comes to her rescue, but not before she has had time to contemplate an end "more terrible than death by fire or torture, for is it not written in the book of the tribe that there is no path through water to the happy hunting-ground," as J. M. Barrie wrote in the novel he made of the original play.

In Ethel Smyth's Cornish opera *The Wreckers*, Thirza is married to the corrupt preacher, Pascoe, who complains that God has not lately guided any ships to be wrecked along the rocky shore so that the locals may pursue their odious custom of plundering their cargo. It emerges that this is because a young fisherman, Mark, has been placing beacon fires on the cliffs to warn ships away. Thirza falls in love with Mark, but the two are caught as they attempt to make their escape. The opera ends with the lovers chained in a cave, left to drown on the rising tide.

Meanwhile, at Execution Dock in Wapping, East London, men guilty of civil offenses committed at sea, such as piracy, were routinely hanged (sometimes using a short rope to prolong the death throes) and then left for three tides before their bodies were removed. This was to serve as a warning to others, although it is also likely that a kind of ritual cleansing was being invoked, as many religious ceremonies involve washing three times.

MEDIEVAL MONKS

Though occasionally the means chosen for gruesome sectarian killings, the tide was more often an inspiration to religious minds. During the medieval period, the study of the tide was kept afloat, if not much advanced, by monastic scholars. The leading English scholar was Robert Grosseteste, who was born in humble circumstances in Suffolk, but rose to high office in the church and at the University of Oxford. As his name suggests, Grosseteste was a man of large frame, which was equaled by his generous demeanor. He was by all accounts amiable, energetic, and inexhaustibly curious, as well as long-lived—in all, an imposing physical and intellectual figure, who bequeathed to his university the beginnings of proper scientific method.

As a young man employed by the bishop of Hereford in the early years of the thirteenth century, Grosseteste found the time to pen a number of essays that betray a consuming interest in physics, as well as commentaries on Aristotle. He wrote about time, optics, the colors of the rainbow, the nature of sound, and the movement of the planets. Although he does not seem to have collected measurements as his predecessor Bede did, Grosseteste thought deeply about the theory of the tides. His work *Questio de fluxu et refluxu maris* ("Inquiry into Ebb and Flow of the Sea") embraces Abu Ma'shar's eight causes of the tides, including factors such as the moon's nearness to the earth, its angle in the sky, and the contribution of the sun.

Grosseteste's greatest preoccupation was with the study of light, which he believed was a divine diffusion as well as a physical phenomenon, and so it is not surprising that he also repeated Abu Ma'shar's belief that the tides were connected with the brightness of the moon. The theory was that the light of the moon was accompanied, like that of the sun, by a heating effect that caused the water of the oceans to expand. Grosseteste also attempted to deal with the obvious flaw in this idea, which was that the tide was high twice a day—once when the moon was above the meridian, but also when it was far below the meridian and the heating effect of its light presumably could not be felt on the water. He solved this conundrum by means of an elegant optical trick in which he supposed that the light of the moon is reflected within the firmament of the sky and is focused to a point in the opposite hemisphere, where it creates a kind of virtual moon with light but no material substance.

Grosseteste is today less remembered than his probable protégé, Roger Bacon, the friar who has often been acclaimed as the progenitor of modern scientific method. In fact, Bacon's methods were little different from those of leading peers in Paris and Oxford. However, his connections were better, and he was not shy about using them. In 1265, he was commissioned by Pope Clement IV, whom he had known before his election, to write a treatise on the place of philosophy in Christian theology. He included a discussion of many areas of science into which he lifted Grosseteste's work on the tides. Bacon shamelessly wrote to the pope, "I have explained one of the most famous, greatest and most difficult effects which are met with in reality, I mean to say the flux and reflux of the sea."

Of course, we know now that Grosseteste's hypothesis was mistaken, and even at the time, its empirical rationale was incomplete, since the highest tides occur at the new moon, when the night sky is at its darkest, as well as at the full moon. Grosseteste also followed Abu Ma'shar in his belief that the winds were a generator of tides, which is true in the local sense that a strong wind can drive water in or out of an estuary, for example, but is not part of the fundamental explanation, which is solely down to yet-to-be-understood forces of gravitation.

In the same year that Bacon offered up his rehash of Grosseteste's theories, Clement IV also appointed one Thomas Aquinas to a position as a theologian in Rome. Aquinas was born in 1225 in the Lazio region of Italy. While his older brothers went off to serve the Holy Roman emperor and king of Sicily, Frederick II, in his battles against papal authority, Thomas took holy orders and became a Dominican friar. The Dominicans did much to revive classical knowledge, and the monastic environment was well fitted to Aquinas's own mission to bring the science of Aristotle (whom he acknowledged simply as "The Philosopher") into Christian theology. In this way, Aquinas believed, baffling natural phenomena might be explained by rational means rather than left as potential evidence against the existence of God.

Under the patronage of Clement IV, the philosopher-saint produced the work for which he is chiefly remembered, the *Summa theologica*, which contains a fivefold argument for the existence of God, an argument purposely grounded in philosophy rather than theology, designed to appeal well beyond the cloister. Such was Aquinas's success that he was canonized not long after his death, despite having no demonstrable miracles to bolster his case.

The first of the five postulates was based on motion. Aquinas believed that spontaneous movement in nature was evidence of the existence of God. Following Aristotle, he reasoned that if an object cannot move of itself, it must have a "mover." One object may be moved by another, and so on, but at the end of the chain, there must be an "unmoved mover," which is God.

The tides naturally warranted closer consideration as an unignorably massive example of such an apparently spontaneous motion. In a later work, called *De occultis operationibus naturae* ("On the Hidden Operations of Nature"), Aquinas aimed to show that a whole range of inexplicable forces that people might think demonic in origin, including magnetism as well as the tides, were in fact responses to natural influences and therefore under the safe control of God. Although Aquinas did not take the theory of Abu Ma'shar and Grosseteste very much further, he did make the intuitive decision to refer to the influence of the moon upon the sea for the first time as a "force."

These influential thinkers helped to ensure that the tides were not completely neglected during the medieval period, and that they might be regarded as a natural phenomenon deserving of an explanation. But they lacked the thorough observations and empirical methods of testing their ideas to make real progress. They were stymied, furthermore, by their adherence to old ideas, such as the belief in the Aristotelian four elements. Here, the astrological identification of the sun with the element of fire and the moon with the element of water led to inevitable confusion in analyzing the influence of each upon the oceans. Observations of the plumpness of sea urchins and shellfish, as well as high tides, at full moon were not unreasonably linked with the wet principle and the moon, for example. In this context, the sun, which caused these fruits of the sea to shrivel in its heat, could never be properly understood as an additional cause of the tides. The occasional correct idea also had to be disentangled from still-current notions of the tides being caused by undersea springs or the breathing of the earth. A more sophisticated understanding would have to wait for the dawn of a new scientific age and the arrival of great scientists such as Galileo Galilei and Isaac Newton.

THE QUEEN'S GUIDE

I am glad that my crossing of Morecambe Bay will not only be accompanied by the RNLI, but also led by the Queen's Guide to the Sands, Cedric Robinson. The position is said to have been brought into existence early in the thirteenth century by King John. The king, of course, had some need of such guides; it was he who lost the crown jewels while escaping from Bishop's Lynn (now King's Lynn) across the marshy wastes of the Wash in Norfolk at the end of his disastrous reign in the autumn of 1216. More reliable records confirm that there have been guides to the sands appointed by the Duchy of Lancaster since at least 1548, when one Thomas Hogeson was made "keeper, conductor and governor" of the Kent sands, following the dissolution of the monasteries. Before this, guides were provided by Cartmel Priory. Today, the position comes with an annual salary of £15 and a grace-and-favor property overlooking the bay, known as Guides Farm.

People's need to cross the sands was sharply reduced when the Furness railway was built in 1857. Today, the Queen's Guide escorts groups doing the walk in order to raise money for charity. There is a busy summer schedule of bookings by hospices, wildlife charities, and sea rescue organizations. The walks are bunched fortnightly when the low spring tides come. The walk I have chosen to join is for the RNLI. Each year, the RNLI rescues three or four hundred people who have become cut off by the tide one way or another around Britain's coasts. More are stranded or run aground in boats, while the phenomenon often misleadingly termed "rip" currents—usually just an ordinary tidal current running close to a beach—is the leading cause of lifeguard callouts. Most of these incidents would be avoided if people knew a bit about the state of the tide when they set out.

Cedric is marking fifty years in service as the Queen's Guide. (The queen has not actually required his services, although in 1985 he did lead her husband, the Duke of Edinburgh, across the bay in a cavalcade of horse-drawn carriages.) "They rung from the tourist office," he tells me excitedly when I first call. "I've been nominated for an award, and—they shouldn't have told me this—I've won it. A black-tie do at the Midland Hotel. So that'll be grand, won't it."

Occasionally, those thinking of making the crossing with no knowledge of the tide have to be set straight. Maps are powerfully suggestive and, it seems, even in this safety-conscious age, can plant in people's minds the idea that they have an absolute right to make any journey that advertises itself as possible on paper. In addition to the unhelpful indication of a "byway open to all vehicular traffic," a box of text warns: "Public Rights of Way across Morecambe Bay can be Dangerous—seek local guidance." But not all do. "I went down one day from Kents Bank station, and behind us two Range Rovers followed," Cedric tells me. "They thought they were going across on a roadway to Morecambe. I said if you set off now you'll never be seen again."

Each guided walk must be carefully planned. A safe path must be reconnoitered afresh each time because a single tide can shift the sandbanks in the channels. Cedric goes out on the sands alone on the low tide before the walk, with the tide times written on the back of his hand, plotting the best route through the channels where they are shallow, and skirting the quicksands. The dry parts of the route he covers quickly by tractor, but he must feel barefoot for a solid path through the deep channels. He then marks the chosen path with laurel branches known as "brobs," which he drives deep into holes made in the sand by a long crowbar. Although the leaves are dead, they stay on the branch, which makes them conspicuous at a distance. On the low tide following this staking out, though, the river can still have moved by more than 100 meters from where the markers were planted.

I JOIN THE WALKERS beginning to gather on the little stone pier at Arnside. It is now raining and a chilly 9°C. There is a buzz of anticipation as we chatter, trying to find the best in the weather: "At least it's not windy." A few passersby sidle up to the organizers, saying that it is something they have always wanted to do. In fact, there is room for them to join the expedition if they wish, but they find new excuses and move off. In the end, we are more than three hundred people, plus a few excited dogs. I start to wonder how exposed to the elements you can truly feel as part of such a large group. A suspicion dawns that the walk must go ahead simply because so many are expecting to do it. But Cedric reassures us that this is not so. The previous year he was forced to cancel four walks on safety grounds. "There's only one person to make their mind up, and that's Cedric," says Cedric, who occasionally refers to himself in the third person.

Our destination is Kents Bank just to the east of Grange-over-Sands, which is clearly visible across the estuary. As the gull flies, it is no more than 5 kilometers, but we have been advised that the walk, avoiding areas of quicksand and fording the channels where they are not too deep, will be anything up to 15 kilometers. We set off along the high-water line down the estuary, past a sign that says:

DANGER

BEWARE

FAST RISING TIDES

QUICKSANDS

HIDDEN CHANNELS

SIRENS WARN OF INCOMING TIDES

IN EMERGENCY PHONE 999 AND ASK FOR COASTGUARD

I pause at the irony that sirens are used to warn people of danger when it is traditionally the case that it is sirens who lure us into it.

The main channel runs close to our shore. The tide is still ebbing, and the water is smooth and murky and of unknowable depth. The banks lie so flat, covered with a shining layer of rainwater, that it is hard to tell where waters end and sands begin. We walk a couple of kilometers overland until we find ourselves on a small, rocky bay, where we stop, looking out to sea. This is where the real walk begins. Now, a mildly transgressive mood seizes the group. A normal walk will often culminate on a headland or a beach with a view out over the sea; the view is the reward. But our walk has just begun.

Cedric is a large, vigorous man with a face so flushed with pink that you would not guess his eighty years, even though white hair is creeping out from under his fluorescent hood. There is no question of frailty or worry about his leadership of this expedition. He gives the signal and, like an army making a charge, we stride out purposefully on a broad front across the marshy grass and then onto the mud and sand. The terrain, if that word can be applied to land that is half the time under the sea, is uniformly marked by the waves. A small-scale ripple pattern in the sand is crosscut with a larger pattern, creating an effect that looks like fish fillets laid out on a slab. Perhaps it was waves of different sizes running in different directions raised by the tide and the wind that wove this complicated texture in the night. Only the occasional cast of a lugworm interrupts the pattern. Elsewhere I find that the tide has run out differently, perhaps at a different angle to the wind, or across sand of a different grain size, and the surface is different. In some places, the sand has been left with a chaotic pattern of anvil-shaped crazy paving; in others, it is smooth as leather.

Every now and then, Cedric brings us to a halt using his whistle, to allow stragglers to catch up. We are such a large flock that I cannot help being reminded of sheepdog trials. After a kilometer or so of walking out across the sand, we reach the present tide line, the limit to which the ebb has pulled back on the edge of the River Kent, where Cedric has planted his laurels, one on the near bank, another a couple of hundred meters away on the far bank. As well as being eminently practical for the job, this plant is appropriate here for another reason, as I learn later. In classical mythology, the nymph Daphne, the child of the river, attempts to flee from the amorous Apollo and, when she knows she cannot escape, cries out for help to her father, who transforms her into a tree. Her arms sprout laurel branches, and her feet root themselves to the spot. Defeated in love, Apollo vows he will tend her as a tree, and he renders her immortal, which is why the laurel's leaves do not drop.

The laurel branches are essential. There are no landmarks out here, and the cues for local orientation are incredibly subtle—the presence or not of a particular seaweed, the slope of a sandbank, the size of the cockleshells. Each of these is unreliable, subject to change with each new tide.

The tide is now nearly slack; the channel is as shallow as it is going to get. As we wade into the water, we are instructed to go not in single file, but to advance abreast in a swath so that followers do not find the sand dangerously softened by the tread of those in the lead. In mid-channel are two more withies, marking areas of dangerous quicksand that we are told to avoid. If we should blunder into quicksand, the advice is counterintuitive but simple: walk faster.

The channel water comes up to my thigh. I taste it and find it is only brackish, warm seawater heavily diluted by cold rain. I can feel the tide still running out against my legs as I wade across. I'm told that the disorienting sensation of watching the water slide smoothly and imperturbably past your immersed body can lead to the panic-inducing illusion that you are not, in fact, walking but swimming, and that this can be avoided only by keeping your sight focused on a distant landmark.

ONE OF OUR GROUP draws the inevitable parallel with the crossing of the Red Sea. Frankly, I am surprised it has taken this long. The most plausible modern explanations of the story of Moses "parting" the Red Sea when he leads the Israelites out of Egypt do indeed center on his superior understanding of the tides in the area, but many questions remain unanswered.

God instructs Moses, "Lift up your rod, and stretch out your hand over the sea and divide it. And the children of Israel shall go on dry ground through the midst of the sea" (Exodus 14:16). Despite the many geographical clues given in the Bible, the exact location where the crossing might have been made is still a matter of speculation. The traditional route involves crossing the top of the Red Sea near the location of the Great Bitter Lake halfway along where the Suez Canal is situated today. But an alternative has the Israelites skirting the edge of the Mediterranean Sea around Lake Bardawil.

In contrast to the Mediterranean, parts of the Red Sea have substantial tides. The tidal range at Suez, at the Red Sea end of the Suez Canal, is up to 2 meters, whereas it is only about a quarter of a meter at Port Said, where the canal reaches the Mediterranean. In biblical times, the sea level was higher, and the tides may have been greater. They will have been augmented as they are now by natural oscillation of the water in the long basins of the Red Sea and the Gulf of Suez at its head, as we have seen occurs in the Adriatic Sea, for example.

Moses is clearly portrayed as a man with a plan, albeit one closely guided by God. Surely he did not count on the unlikely fluke of a tsunami, such as the one caused by the volcanic destruction of the Greek island of Santorini, which has been advanced as an alternative explanation of events. No. From his time spent in the wilderness, Moses would have been familiar with Red Sea shores and their considerable tides, and it is clear from the biblical account that timing is an important feature of the Israelites' journey. The Bible says nothing explicit about the tide, but it does say, "The Lord caused the sea to go back by a strong east wind *all that night*" (my italics). Sustained winds can indeed produce surprisingly large changes in sea level. In this location, they could have driven the level down by more than 2 meters and pushed the shore out to sea by as much as a kilometer, according to some calculations. If a strong wind could have helped to drive the local sea level down at the site of the crossing, its abrupt cessation would equally have allowed the sea to run back in, perhaps in a matter of minutes.

There is also a possible clue here to the nature of the tide. Although Suez today has normal tides twice a day, some locations at the south end of the Red Sea have strongly diurnal tides, with just one high and one low in each twenty-four-hour period, because of the oscillating basin effect of the Red Sea. With the sea level higher than it is today, it is conceivable that there was then also a stronger diurnal tide at the north end of the sea, which, in conjunction with the east wind, might have caused a night-long period of low water, allowing Moses and his people to cross before the flood tide returned in the morning and drowned the pursuing Egyptians.

Other biblical details still confound this rationale and many others, however. It is said that Moses led six hundred thousand people and their livestock to safety, and there is no chance that such an army could have made the crossing in the space of a single tide. The powerful image of the waters dividing—"the waters were a wall to them on their right hand and their left"—is even more problematic, suggesting the sudden emergence of some kind of causeway, which is an unlikely bathymetric feature. Perhaps we should not be so literal. The real marvel is in the storytelling.

IN MORECAMBE BAY, the scene is resolutely horizontal. Flat water, flat sand, in all directions. But this bland vision belies subtle differences. Beneath my feet I can feel a change. On the far side of the channel, the sand is muddier, and here and there are patches where it seems to have turned to a kind of jelly. Only the surface tension of the water saturating the sand produces the slightest skin on top, giving a dangerous impression of solidity. It is like walking across a barely set dessert. If you rest your foot for long on one of these patches, I find, it is sucked hard into it. If you try to push free with your other foot, it soon becomes stuck too. Some unlucky souls may have read the signs and thought "quicksands" were a fabulous exaggeration, something only ever found in adventure films. But they are not. They have been known to swallow cars and tractors. The danger for walkers is not so much that you may be sucked under completely, but that you prove unable to free yourself before the tide comes in. On one occasion, Cedric was helping a television crew demonstrate the dangers of quicksand. Volunteers were understandably reluctant to come forward, so a human dummy was constructed, which duly settled into the sand and could not be retrieved. For weeks afterward, people on the shore reported that there was a man drowning in the bay, until one day Cedric went out with a saw and hacked the "body" in two and retrieved the top half of the torso so it would not give rise to distress.

The habitat underfoot is changing too. Now there are abundant cockleshells strewn in irregular bands across the surface of the sand. It is the wrong time of year for the cockles themselves, though, and so we see none of the birds that come to feed on them. When he sees something of interest, I'm told—a large jellyfish, or a ray caught in the shallows—Cedric will stop to point it out before placing it in deeper water to watch how it swims off. Most often, it is the small flounders known as flukes. But today we see no fish. All the life is below the surface of the sand.

We are now a couple of kilometers from the nearest land. It feels strange to be here. The ground is firm underfoot, but the surrounding expanse is mirror bright. We can see our reflections whole, as if standing miraculously on the surface of a pond. The skim coat of water on the sand echoes the color of the sky. Only on the horizon is there an ominous darker band of more bad weather on the way. We have walked about 7 kilometers out to the point where we crossed the Kent. Now it is time to double back along the tidal flats on the Grange side of the estuary.

Back on terra firma, we tender our thanks to Cedric and hop on the bus that will now take us the long way around back to our cars. It has been a walk like no other—easy going because so flat, and featureless in some ways, but also a trespass into an alien environment, like walking on the surface of a different planet. As I leave the bay, the tide is yet to return. It is hard to believe that this place (for, to us now, this temporary land that is often sea is unquestionably a place) was under several meters of water six hours ago, and will be so again in a few hours' time. It is only when I see it at last at high tide that it strikes me what a very odd thing we have done—a madness and an impossibility now.

. . .

LATER, I HEAR CEDRIC'S STORY. Cedric began working on the sands as a boy with his father, catching flounder and, in summer, skimming the inch-deep water at low tide for shrimp by night, while in winter, digging for cockles. A horse and cart brought them out onto the sands and took in what they managed to catch. Later, he began to be approached for local advice by commercial interests ambitious to conquer the nuisance to modernity that they found Morecambe Bay to be: Would it be feasible to build a barrage? Where would it be best to lay electricity cables? From these beginnings, he slid into the more congenial role of Queen's Guide to the Sands. The title means little to him, he insists. "I don't reckon meself anything. I've been on the sands all me life and I love the sands, that's all." It is Cedric who has made the job relevant again. By now, he has had a hand in raising tens of millions of pounds for charities.

But he enjoys being out on the sands alone most of all. "It's such a vast area, and it's never the same twice. Especially at sunrise, coming back to see the houses rising, and the cows in for milking."

The bay is more dangerous now than it used to be, he tells me. The rivers are deeper, and the sands shift around more. All the dangers that have claimed lives in the past are still present: the quicksand, the fog and heavy rain that obscure the views of the surrounding hills essential for orientation, the tricks your eyes can play so that what seems to be a distant mirage turns out to be a glittering wave front advancing toward you at a gallop. And always the tide.

5

TERRA
INFIRMA

GALILEO'S TROUBLE

Venice is one of the last places you might think of as having a problem of water supply. But for centuries, until a pipeline was finally laid in 1884, the city was reliant on freshwater brought across the lagoon from rivers on the mainland in massive barges.

It was quite likely on one of these water barges that the young physicist Galileo Galilei arrived in the "Most Serene Republic" in the spring of 1592. Encouraged by his sponsor, the marquess Guidobaldo del Monte, he resigned from the University of Pisa, where, though appointed to the chair of mathematics, he had always struggled to gain the support he needed for his work. He left his hometown with a head full of observations and ideas. It is probable that he never actually dropped different weights from Pisa's famous leaning tower, as legend has it, but he certainly did have ample time in the cathedral there to observe how the lanterns always swung back and forth at exactly the same rate, no matter what the strength of the breeze outside that set them going.

He spent that summer in Venice with Guidobaldo, an enthusiast of mechanics, doing experiments on the motion of projectiles, before taking up a new appointment at Padua, a university larger and more prestigious than Pisa, and one where, under the governance of the liberal Venetian Republic, he would have fewer difficulties with the authorities. At Padua, you can still see the lectern at which he taught, a surprisingly rough-hewn set of steps up to a chunky and unadorned wooden platform.

Galileo was not a Copernican when he left Pisa—he had not seen for himself any compelling evidence to support the idea that the earth rotates around the sun—but by the time he left Venice he was. What did he see there that made him change his mind? In the first place, his investigation of projectiles indicated to him that bodies could move under the influence of more than one force at a time—in this case, the propulsive force of the explosion to launch a missile and the still unrecognized force of gravity pulling it off its path and back to earth. But second, there was the persistent impression of those water barge journeys. He had seen how their vital cargo lay still when the boat was proceeding at a steady speed but would slop about when the boat changed speed or direction. When a barge slowed as it came in to dock, for example, its load of water would rise up in the bow and fall in the stern. From this behavior of the water in the barge, Galileo was able to deduce an important general law of physics: that without an external viewpoint, a body in constant motion is indistinguishable from one at rest. As he finally left Venice for Padua, no doubt again on such a barge, this time emptied of its precious load, passing along the Brenta Canal in front of villas newly completed by the architect Andrea Palladio, Galileo had the germ of his "two world systems" in his head, although the troublesome treatise would not be published until forty years later, after many personal trials.

The water in the barge gave Galileo a more specific clue to a very earthbound phenomenon that might offer the key to this cosmic question. After it had run forward when the barge stopped, the water would then oscillate back and forth for some time at a steady rate like the Pisa cathedral lanterns. It occurred to Galileo that the whole Adriatic Sea might be similarly affected by an external force that would explain its tidal movements. This was his crucial insight: what if the whole earth moved?

The Adriatic is a long, roughly rectangular body of water enclosed at the north end by the Gulf of Venice and nearly enclosed at the south end by the Strait of Otranto. Galileo knew that the tides at each end of the sea, in ports such as Brindisi and Trieste, as well as in Venice itself, were large by Mediterranean standards, while on the coast halfway along, at places like Pescara and Ancona, there was hardly any tidal movement. The water barge modeled the situation exactly. The level of the freshwater at the aft end of the tank in which it was being transported fell when it rose at the forward end, and vice versa. A point halfway along the tank seemed like a kind of fulcrum where the level did not alter.

Not only was the water barge a model for the Adriatic, but the Adriatic was a model for the world's oceans, for, as the Roman geographer Strabo had long ago observed, "Here are almost the only parts of Our Sea [the Mediterranean] that behave like the ocean, and both the ebb-tides and the flood-tides produced here are similar to those of the ocean, since by them the greater part of the plain is made full of lagoons."

Galileo would have been familiar with the idea of bodies having resonant frequencies governed by their size from his father Vincenzo, a lutenist and an important music theorist of the day. The water in the barge might slosh back and forth every few seconds, but the period of oscillation of the Adriatic Sea is observed to be about twenty-two hours, or near enough to a day.

This sketchy explanation of the earth's tides would become the linchpin of Galileo's argument in favor of the heliocentric solar system. There was no convincing astronomical evidence that the earth moved around the sun; the best on offer was the apparent passage of sunspots across the face of the sun, which was and remains an area of controversy. Falling bodies, projectiles, and pendulums provided limited experimental evidence, but the idea of oceanic tides forced to move by the motion of the earth itself was more compelling still. Soon after he was ensconced at Padua, Galileo bragged to his German rival Johannes Kepler that he had been able to use observations of natural phenomena to produce irrefutable evidence in favor of the Copernican theory. Kepler guessed he meant the tides. But it was not until many years later that Galileo publicly revealed his theory of the tides as part of his controversial proof that the earth moves.

The *Dialogue Concerning the Two Chief World Systems*, the treatise that Galileo finally was able to publish in 1632, compares the geocentric planetary scheme of Ptolemy and the heliocentric theory of Copernicus. The frontispiece of the work shows Aristotle, Ptolemy, and Copernicus, each in appropriate period dress, standing in contemplation on the seashore at Livorno, where the tidal range happens to be less than a foot. In this work, Galileo stated his conviction that closer study of the movement of the oceans would furnish proof of the new theory that the earth revolved around the sun. He believed that the daily rotation of the earth upon its axis and its annual orbit around the sun combined to produce complex patterns of planetary acceleration and deceleration, and that one way in which these forces manifested themselves was in the motion back and forth of the waters of the earth's oceans in their respective basins.

The moon played no part in Galileo's theory, despite observations since antiquity that the strength of the tides seemed to be linked to the phases of the moon. This omission placed him at odds not only with classical philosophers such as Aristotle, but also with Kepler, who was convinced that the moon was somehow responsible but could not yet work out how. Unlike these great men, who were forced to rely for their theories on astronomical measurements—astronomical, and therefore remote and unverifiable—Galileo felt he was in a position to offer the first empirical evidence to explain the phenomenon of the tides. So cherished was this hope that Galileo had first chosen "Dialogue on the Tides" as the working title of his groundbreaking treatise, and only at the insistence of the Inquisition did he amend it to its more enigmatic final form.

Galileo knew, of course, that most places experience two tides a day, not one as his calculations predicted and as was nearly the case in the anomalous Adriatic. He tried to develop his theory to allow for this major discrepancy, suggesting that twice-daily tides arose as a kind of echo effect (although he was bolstered, too, in his mistaken core assumption by anecdotal reports that here and there were indeed other places that experienced only one tide a day). By the 1610s, Galileo's growing confidence in Copernicus's theory led him into dispute with the Catholic Church. In earlier correspondence with Rome, he had judiciously omitted any discussion of the tides, even though this was key to his argument in favor of Copernicanism. Thinking it might be best to reason directly with his critics, the astronomer traveled to Rome in December 1615, believing that his rationalization of the tides provided the tangible evidence of the earth's motion around the sun that the church had demanded to see from him. Yet the meeting went disastrously wrong for Galileo, and for science. Pope Paul V reacted by setting up a commission that formally declared Copernicanism to be heretical. In February 1616, the pope's foremost cardinal, Roberto Bellarmino, summoned Galileo and told him in front of representatives of the Inquisition that he must not promote his theory or even believe it.

But Galileo continued to think about the problem. A new difficulty now presented itself, in that his empirical evidence seemed to be growing weaker. Tidal data solicited from various parts of the world confirmed that the tide was generally high *twice* a day, and other scientists also pointed out his basic error. Nevertheless, Galileo continued to hold that his water barge theory had universal validity. In the North Atlantic, from where he had requested tidal measurements along various coasts, he argued, it just so happened that the natural period of oscillation of the whole ocean was about twelve hours. (It seems odd that a scientist who spent so long observing the motion of pendulums should believe that a vast ocean would slop back and forth *faster* than a small sea such as the Adriatic.) And Galileo, of course, persisted in secretly believing in the main thing, which was that the earth moves, and that the tides would somehow be fundamental in proving it. "Copernicanism mattered to Galileo, and the reasons for this were not simply scientific," as David Wootton put it in a recent biography. Albert Einstein later wrote, forgivingly, that Galileo's heart had ruled his head in teasing out his muddled rationale for the tides.

In 1621, both Pope Paul V and Cardinal Bellarmino died. Feeling that the injunction against him had lapsed with their passing, Galileo incautiously began to revise his discourse on tides, preparing to reopen the matter of Copernicanism, although there was still no immediate prospect of his being able to publish. It was not until 1630 that Galileo finally set out for Rome with the completed manuscript of the "two world systems" and was able to negotiate publication, which followed, after further delays owing to an outbreak of plague, in 1632.

This should have been Galileo's long-awaited moment of triumph. But he had yet other enemies, and those men pointed out fresh aspects in the dialogue where the astronomer could be accused of heresy. In the trial before the Inquisition that followed in April 1633, the inevitable guilty verdict was delivered and the great scientist forced to confess his "error," although the sentence of life imprisonment that was passed was soon commuted to relatively comfortable house arrest.

More important for Galileo was that he was able to continue his research, although his next major work, on mechanics and statics, *Two New Sciences*, he took care to smuggle out of Italy to be published in Leiden. Now in his seventies, he continued to improve his tidal theory, attempting to correlate the small variations in daily, monthly, and annual cycles found in both lunar position and the tides, and still hoping to clinch his proof of the motion of the earth.

Clearly, Galileo was a stubborn man, but he was stubborn with good cause. He had what he felt was a sound theory of the earth's motion and solid evidence for it from the nature of the tides. More significant still was his modern scientific drive to be able to explain natural phenomena in principle by means of purely mathematical deduction. It was his misfortune that this unshakable conviction appeared to challenge the teachings of the church.

As Wootton writes, Galileo was, in the end, "prepared to gamble everything on his theory of the tides, even when it was seriously incomplete." It is a moot point, though, whether Galileo would have avoided papal proscription had his theory been substantially complete or correct. Wrong as he was, Galileo's dogged work on tides did nevertheless yield some valuable observations, including confirming that the modest tides at Venice and at the southern end of the Adriatic Sea are nevertheless greater than those at the midpoints along its coast, and that seawater, acted upon by whatever great force, can indeed be made to move in the same way as water slopping in a bath. Moreover, Galileo's insistence on collecting proper data, rather than merely conjecturing, was a signal of the scientific revolution, even if, in trying to fit the data to the problem rather than the other way around, Galileo ultimately allowed himself to be deceived.

The glaring flaw in Galileo's theory was that it had no role for the moon. For centuries, people the world over had noticed there was a link between the phase of the moon and its position in the sky, and the heights and times of the tides, but they had been unable to explain it. The first modern scientist had done no better by insisting that it was just the rotation of the planet causing the seas to move about. His difficulties were not helped by the fact that he drew much of his data from the Adriatic, where nontidal oscillations caused, for example, by strong winds or differences in atmospheric pressure, at a natural period of twenty-two hours, often swamp the modest Mediterranean lunar tide. Put simply, Galileo was unable to think outside the box, and his box was the Adriatic Sea.

ACQUA ALTA

VENICE

	HW	LW	HW	LW
November 3, 1966	01:48	06:20	12:11	19:55
	0.7 m	0.5 m	0.8 m	0.1 m
November 4, 1966	03:30	07:30	12:48	21:07
	0.6 m	0.6 m	0.7 m	0.2 m

Nearly thirty years ago, I traveled to Venice to write about plans to build flood barriers across the three narrow inlets connecting the lagoon that encircles the city with the Adriatic Sea beyond. It was then already fully two decades since the flood of November 4, 1966, which greatly damaged most of the buildings and much of the priceless artwork in the city, and no protective measures were yet in place to prevent a recurrence.

A prolonged spell of extraordinarily heavy rain over much of northern Italy provided the prelude to the highest storm surge ever recorded in the city's long history. The water level in the lagoon began to rise around noon on the third of November and reached its peak at midnight. All through the next day, the waters continued to rise, driven on by sea waves growing ever more violent. Around noon, the waves began to weaken, but the sea level in Venice continued to rise, reaching a new peak in the evening, an *acqua alta* more than half a meter higher than any previously recorded. The tide gauge at the Punta della Salute on the Grand Canal read 1.94 meters (an average high tide reaches about 80 centimeters). St. Mark's Square lay under more than a meter of water, and five thousand people lost their homes.

It is usually said that Venice flooded owing to a fatal combination of a high tide, freshwater flowing into the lagoon from rivers swollen by torrential rains, and a strong sirocco wind blowing northward, pushing seawater before it up the length of the Adriatic. In fact, the floodwaters might have reached considerably higher than they did were it not for the fact that the storm surge was actually at its highest when the tide was halfway out. The effect of the tide was to prolong rather than to intensify the flood. The prediction for the tide alone at Venice on that day was, in fact, for a relatively small tide, which would remain high from the wee hours all the way through until dusk. This meant that the water level, swollen by the nighttime rains, did not recede as usual, but remained high for a full twenty-four hours. It was this prolonged exposure to the water and waves that caused the most devastation, with three-quarters of the city's buildings and the homes and businesses they contained destroyed or damaged.

Venice has always experienced floods, but these grew more frequent during the twentieth century. Today, parts of the city are flooded more than forty times a year. It is routine for hotels to direct their visitors to a website that warns of coming *acque alte* and provides a map of the raised walkways that are laid out across the city so that some semblance of normality continues.

It is not only the combination of storms and tides that places the city at risk, but human activities, such as large-scale construction and the extraction of drinking water from aquifers under the city, which was not stopped until the 1970s. The sum of these effects is that a tide recorded in Venice now is 250 millimeters higher than the equivalent tide in the year 1900. (Today, it is vast cities such as Bangkok, where the land is sinking by 13 millimeters a year, that are in greater danger from this shortsighted rapaciousness.) It is difficult, but important, for scientists to know how much of Venice's "sinking" is due to local subsidence of the land and how much is due to rising sea levels, and separating the two calls for some imaginative thinking.

Dario Camuffo of the Italian National Research Council in Padua looked at the "photographic" paintings made by Antonio Canaletto and his pupils with the aid of a camera obscura to find proxy sea-level data from the eighteenth century, before the city had a tide gauge.

By comparing the height of the waterline band of algae on various buildings then and now, he was able to confirm that the sea level rose at a constant rate of 1.9 millimeters a year for at least the couple of centuries before about 1930, when the rate increased to 2.5 millimeters a year, owing, it is thought, to the impact of water extraction and other modern intrusions. Camuffo recently extended his historical data back another 150 years or so by examining palazzi that appear in the backgrounds of paintings by the sixteenth-century Venetian artist Paolo Veronese.

After the 1966 flood, it was clear that the city needed a barrier, or rather, a barrier for each of the lagoon entrances—one at the town of Chioggia to the west, another at the east end of Lido island, and a third at Malamocco in the middle, which is the entrance to the lagoon used by oil tankers and other ships to reach the port and refineries of Mestre, Venice's unlovely twin town on *Terra Firma*, as the Venetians long ago named the mainland part of the republic.

The precise form the barrier design should take became the subject of intense debate. Environmental lobbyists expressed concern that permanent barriers would damage the delicate tidal ecosystem of the lagoon. Large quantities of soil, sewage, and agricultural chemicals are washed into the lagoon, and natural tidal flushing is vital for their dispersal. Even a few days' closure of the lagoon each year, it was felt, could lead to accumulations that would do permanent damage to marine life. Cruise ship operators and shipping companies working out of the port at Mestre, on the other hand, sought assurances that their huge vessels would still be able to use the lagoon. The lagoon's long-established fish farmers feared for their livelihoods too. By 1973, the Italian government was forced to accept that any solution would have to guarantee the "preservation of the ecological and physical unity of the lagoon."

This meant that the barriers would have to be designed in such a way that they would have no effect on tidal currents most of the time, and would come into operation to provide a temporary seawall when an *acqua alta* was forecast. Numerous proposals, of varying degrees of suitability, were submitted by Italian engineering and construction companies, many of them gaily ignoring the conditions set out in the brief. One of the more plausible schemes came from the tire company Pirelli. Its plan called for giant inflatable sausages up to a kilometer in length made of rubberized polyester. The idea was that the tubes would lie flat and empty on the seabed until pumped full of air, when they would rise up to block each of the lagoon's entrances.

It was February when I made my journey, and I remember being astounded that even in this foggy season my vaporetto was full of what seemed to be vacation revelers. Later, I realized why this was so as, puffed up with the importance of my project, I threaded my way impatiently to interviews through masked crowds gathering for the annual *Carnevale*. My principal appointment was with the technical director of Consorzio Venezia Nuova, the consortium of engineering and construction businesses newly constituted under the authority of the Venice water board, the Magistrato alle Acque, to address the challenge of saving the city. On November 4, 1986, twenty years to the day after the floods, the Italian prime minister, Bettino Craxi, had announced a 6,000-billion-lire award ($4.2 billion, and including monies already spent) to take the project to completion for operation by 1996, now based around a final design for a mobile steel barrier called MOSE, Italian for "Moses," who famously saved the Israelites by holding back the Red Sea, and also a strained acronym for MOdulo Sperimentale Elettromeccanico. I wanted to see exactly how the money was going to be spent.

By the time of my visit in 1988, the consortium had made considerable strides. In various test facilities across Europe, models of parts of the lagoon and the barrier were being evaluated. Meanwhile, work was already proceeding on dredging the channels and building up the *barene*, or salt marshes and other natural defenses within the lagoon, as a buffer against the effects of future high tides. I spoke to engineers and scientists at universities and companies in the Netherlands, Denmark, and Britain involved in the project. I learned about the serious consideration theoreticians were giving to the effects of the altered flow of water between the lagoon and the Adriatic. Everybody acknowledged that the final design would have to be more sophisticated than existing schemes such as the Thames Barrier or even the Oosterscheldekering, designed to protect low-lying areas of the Netherlands.

Finally, in an industrial shed in a desolate suburb of Rome, I saw for myself a working model of part of the proposed barrier: a series of metal containers painted bright yellow and shaped like squares of chocolate lying in a large water tank. These were a 1:40 scale model of part of one of the barriers. I observed how the containers lay flat on the bottom of the tank when filled with water, and how they then rose to the surface, forming a wall between one end of the tank and the other, when air was pumped in, bobbing gently as the waves broke against them. In all, seventy-eight such containers would be placed at the three entrances of the lagoon, resting for most of the time on the seabed in concrete channels, to be swung up into operation when needed by the injection of compressed air. As I looked on, I recalled the words of Augusto Ghetti, a distinguished professor of hydraulics at the University of Padua, whom I had interviewed a few days earlier, and who was gently skeptical about all the complicated machinery involved. "The technicians think there is no problem to introduce the air. I think it is not so easy."

FLOOD ON FLOOD

The tide is a contributing factor to the risk of flooding at Venice, but not the principal factor. Along the coasts of the North Sea, however, the balance of natural influences on the sea level is different, and the tide played a much greater role in the catastrophic North Sea flooding

in 1953 that forced the Dutch and British to take permanent measures to protect their major cities. In the River Thames, for example, a
storm surge can increase the height of the tide by more than a meter,
and the tides have a normal range of 6 meters. (At New Orleans,
by way of contrast, it happens that there is nearly no tide, even less
than in much of the Mediterranean Sea, and the tidal range of 30
centimeters or so is negligible compared to the temporary sea-level
rise produced by severe low-pressure areas such as Hurricane Katrina,
which devastated the city in 2005.)

The story of the 1953 floods along the east coast of England is told
in a book commissioned by Essex County Council called *The Great
Tide*. Hilda Grieve's account is far better than its municipal origin
might lead you to think—gripping in its personal detail, and forensic
in its examination of the reasons why more than three hundred lives
were lost. As was to be seen in Venice thirteen years later, it was the
prolongation of the period of high water, driven on by the wind, that
caused sea defenses to be breached in a sustained fashion and far more
water to flood inland than would have been the case with a short-lived
surge. The exceptional weather factor on this occasion was a vicious
cyclone that developed in northern Britain and then swept down the
North Sea, scouring the coast with a force-ten northwesterly gale.
Here it coincided with a high spring tide. The tide was, of course, predicted, but the rapidly developing weather situation was not. Also it
was a Saturday. The offices of government agencies and local authorities were closed for the weekend, and so even the most basic warnings
from northern points along the coast where the coincidence of tide,
wind, and extreme low pressure was felt first, were not passed south
where the storm would wreak such havoc a few hours later.

It is not always the initial onrush of water that characterizes such
a disaster; the flood tide is expected, after all, in some degree. It is
its failure to ebb again when it should a few hours later that is truly
frightening. That is when coast dwellers might start thinking end
thoughts. What if it never ebbs? What if the next flood simply over-

tops the last? And the next, and the next? Grieve shows that this is an old fear, citing the early chronicler Matthew Paris, who describes the devastation and loss of life along the coasts of East Anglia and even in towns well inland wrought by the Great Martinmas Tide of 1236:

> There burst in astonishing floods of the sea, by night, suddenly, and a most mighty wind resounded, with great and unusual sea and river floods together, which, especially in maritime places, deprived all ports of ships, tearing away their anchors, drowned a multitude of men, destroyed flocks of sheep and herds of cattle, plucked out trees by the roots, overturned dwellings, dispersed beaches. And the ocean rose flowing with increase for two days and one night in between, which is unheard of; nor was there, as by usual custom, ebb and flow, the most mighty violence of the contrary winds, as it is said, preventing.

On this night in 1953—Saturday, January 31—and all during the following day, the buildup of water likewise continued progressively over two tidal cycles. It meant that the ebb, when it did finally come, was both fast and furious, and more destructive than the original flood, dragging whole houses with it back to the sea. An additional factor in the loss of life, according to Grieve, was the ignorance of many of those living closest to the coast, especially those on Canvey Island, who had moved out of London and knew little about the dangers of winds and tides. London itself had a narrow escape; water lapped the tops of the embankments as far upriver as Victoria and Chelsea.

In Norfolk, where I live, the sea breached the defenses first at Hunstanton, and sixty-five people were drowned. Across the county, a 2-meter wall of water overflowed the banks of the River Yare at Great Yarmouth at 8:00 p.m., well before the predicted time of high tide. The wave raced into the Southtown and Cobholm districts of

the town. The Breydon seawall gave way at 11:00 p.m., and thirty-five hundred homes were flooded. The high tide passed down the coast like a traveling wave as Bede's data had long ago showed that it did. At Aldeburgh, flood defenses strengthened as recently as 1949 proved useless. Settlements along the Thames, such as at Canvey Island, were hit in the wee hours. Those not awakened by the noise of the storm were awakened by the water rising into their bedrooms, sometimes leaving them with no escape other than to break a hole through the ceiling and climb into the attic and then onto the roof to wait for rescue.

Across the sea in Holland, always at greater risk, there was at least a basic early-warning system in place, but it did not extend far enough and was not heard by many. In the country's most devastating flood since 1421, more than eighteen hundred people perished.

Much of Dutch life is concerned with keeping the sea at bay. The completion of the 20-mile Afsluitdijk to enclose the Zuiderzee in 1927, and the more recent Delta Works to protect land at the mouths of the Rhine, Meuse, and Scheldt Rivers are important modern initiatives, but the building of dikes in order to safeguard populations living on low rises of land goes back to around 1000 CE. However, it is worth remembering that the high tide has, on occasion, given the country a unique strategic weapon. In the west of the country in 1584, William of Orange ordered that the seawalls be breached so that his rebel forces could relieve the cities of Bruges, Ghent, and Antwerp, then under Spanish occupation. The tactic failed, and 60,000 acres of land were wrecked for agriculture. Further to the east, the *Oude Waterlinie* ("Old Waterline") was an elaborate system of water management stretching inland from the coast just east of Amsterdam, installed in response to sea floods that had devastated several Dutch cities in the 1570s. In 1672, the waterline was put into reverse, and the water let in, to confront a new threat of invasion by Louis XIV. The admitted floodwaters formed a broad barrier along an 80-kilometer north–south defensive line. Just 40 centimeters' depth of seawater, stretching far across the expanse of the fields, was enough to bring the heavily laden French troops to a halt. Soldiers who attempted to advance along the dikes between the fields were easily picked off by the lesser Dutch forces. Only during a short freeze that winter were the French soldiers briefly able to advance across the ice. By the spring, Holland was saved, although the farmland would take years to recover.

The potential for flooding their own territory may have saved the Dutch again during the First World War. Shortly before that war started, Queen Wilhelmina of the Netherlands visited the belligerent German kaiser, Wilhelm II. When he remarked provocatively how much taller his guards were than hers, the queen is said to have replied, "True, your majesty, your guards are seven feet tall. But when we open our dikes, the water is ten feet deep." The Netherlands remained neutral during that war and was not visited by the kaiser's army.

During the Second World War, both sides employed flooding again. In 1944, the retreating Nazis flooded parts of the country in order to impede the Allied advance, while General Eisenhower ordered that the island of Walcheren, strategically located at the mouth of the Western Scheldt River commanding access to the port of Antwerp, be flooded to drive out the Germans garrisoned there. In all, according to Adriaan de Kraker, an assistant professor of historical geography at the Free University of Amsterdam, fully a third of all the floods in parts of the Netherlands since 1500 have been deliberately incurred as defensive measures.

The East Anglian communities that were worst hit by the storm surge do not need reminding of their liminal status. It is something they normally celebrate. For these are edge lands, not just in geography, but socially too. They are the coasts in Britain where the land is sinking the fastest, a corresponding sinking as the island is lifted up in the north, feeling its freedom from the weight of the glacier that once lay upon it. But they are also coasts, especially in Essex and Kent along the lower Thames estuary, that exist as a kind of nether realm where it seems that the writ of law, and planning law especially, does not run. Everything has an improvised appearance. The law that matters more is the law of nature; everybody understands that a salt-marsh shack is more likely to suffer a visit by the tide than by the council inspector. The buildings here are not permanent, because the land is not permanent. It is the expanse and softness of the muddy shore, territory that is sometimes there and sometimes not, that encourages this permissiveness: you do not see the same dishevelment on rocky coasts where the buildings cling primly to their solid foundations. Here, too, humanity's offenses will one day be erased, and so it is that the socially marginalized find their way to this geographical margin.

Do not think this sense of transience is imagined. Many settlements have disappeared beneath the tide. The most famous of these is the town of Dunwich in Suffolk, in medieval times a port to rival London. It seems now almost a legend, like Lyonnesse, or Ys, the

Breton city whose long-ago towers can supposedly be seen only when the sea is exceptionally calm and clear, and which is celebrated in Debussy's piano prelude *La cathédrale engloutie* ("The Sunken Cathedral"). Fabled Ys was founded on dry land, but being gradually threatened by the sea, it erected a protective seawall. One night of a high tide, though, the citizens forgot to close the gate in the wall . . .

Similar tales are told of Cantre'r Gwaelod and Tyno Helig in Wales. According to medieval verses, the former was a domain of no less than sixteen cities that was long ago lost below the waters of Cardigan Bay when its ruler was judged for his sinfulness. The latter, supposedly located between Conwy and Anglesey, suffered a similar fate. Both events, if they happened at all, may stem from a single catastrophic event that has been cautiously dated to the sixth century. Underwater "ruins" found on the Lavan sands in the Menai Strait have been dismissed as natural formations, but they are enough to keep the folktale alive.

But Dunwich was there all right. The ruins lie just off the present-day coastline at a depth of only a few meters, although there is no chance of glimpsing them directly in the soupy, nutrient-rich North Sea. In 2013, David Sear at the University of Southampton and Tim LeBas, a sonar acoustics expert at the National Oceanography Centre, provided the first sight of the city in 750 years. Their murky images show a network of streets and the remains of eight churches—and evidence of defenses that were repeatedly breached during the exceptional storms of the climatic transition that began in the thirteenth century from a warm medieval period to what came to be labeled the Little Ice Age, visiting doom on the once prosperous port.

THE THAMES BARRIER was built in response to the 1953 floods. It became operational in October 1982 and was first called into use four months later, in February 1983. Now, it is used more often than anybody ever anticipated. There were 50 closures during the stormy winter of 2013–14 alone, compared with 124 closures in the whole of the thirty years of the barrier's history before then.

By 2100, water levels in the tidal Thames are expected to rise between 20 and 90 centimeters, owing to global warming. Nonetheless, it is currently estimated that the barrier will remain effective for longer than was first planned, perhaps up until the year 2070 rather than 2030. Factors related to climate change are thought less likely to lead to an increase in storm surge heights and frequencies in the North Sea than they were previously. Nevertheless, planning for a new barrier will need to start by 2050, and planners have tentatively identified a site for a new and higher barrier near Longreach, just upstream of the Dartford road crossing.

In April after the winter of many closures, I join a technical tour of the barrier with a team of drainage engineers from one of the nearby local authorities. Our guide, George, is a retired civil engineer who has been involved with the barrier for most of its working life. He has a pleasingly deadpan delivery and is at pains to point out that he will not be a spokesperson for the Environment Agency, which has responsibility for operating the barrier. He begins by sitting us down to watch the inevitable introduction on DVD. "It's a bit corporate," he apologizes.

The statistics gush by, figures borne on a stream of heroic background music. The barrier comprises nine piers with 61 meters of clearance between them to allow for the passage of shipping. Each of the four main "rising sector" gates weighs 3,200 metric tons and fits between its shiny steel piers to within a tolerance of half a millimeter. The barrier is designed to safeguard London against a one-in-a-thousand-years risk (the 1953 floods were judged to be a one-in-three-hundred-years event).

Afterward, George leads us down and up steel stairways and along pristine concrete tunnels in the riverbed out to one of the midstream piers. Each pier is shaped like an upturned boat and paneled in stainless steel. I am surprised to find they are timber-framed inside. The quirky design was developed at the insistence of the Greater London Council, which saw the merit of creating a structure that might be regarded as iconic as well as functional, and threw in the extra funding to make it possible. We are shown the mechanics of the barrier and are reassured about the many levels of redundancy incorporated to reduce the risk of operational failure. We learn how incoming data—tide predictions from the Admiralty, the volume of water coming downriver that is measured upstream at Kingston, and the height of the surge that might be produced by storm conditions, estimated using data supplied by the Meteorological Office—are combined and assessed before the decision to close the barrier is made. In practice, that call must usually be made before it is certain there will be a storm surge, in order to allow time for key staff to be called in from home, as well as for the machinery to swing into action. The moment when closure begins and the moment when the barrier is reopened once the danger has passed must both be carefully judged. "You can't change your mind, because of the risk of sending a wave front up- or down-river," George explains.

London is better prepared today than it was in 1953, yet also fundamentally more vulnerable. Construction of luxury apartments right up to the riverbanks places hard embankments where once there were absorbent salt marshes, and these leave an incoming storm surge nowhere to go but further upstream. The Environment Agency video promises "better places for future generations." But George is less sanguine. In the event of another flood, "the downstream defenses wouldn't fail so early, and London would suffer."

THE VENICE BARRIER

Twenty-five years on, I return to Venice to catch up on progress on that city's barrier. The city is still without effective protection from the sea. From my airplane window, its most familiar feature, its charming canals, now appear to me like streets flooded after a disaster. People cluster in little squares. Here and there, vessels ply to and forth, as if bringing aid. The in-flight magazine glibly informs me that the "*cognoscenti* prefer Venice in winter. Even the constant threat of the *acqua alta*—the winter floods that inundate St. Mark's Square with increasing regularity—add a little taste of adventure (it's worth packing wellie boots)." Perhaps the fight has been given up.

Floods in November 2012 were the sixth worst since the city began to keep records in 1872. That time, the water at the Punta della Salute tide gauge rose to 1.49 meters. Even the raised duckboards brought out at each *acqua alta* floated away. But the worst flood before that was only as long ago as 2008, which gives a clue to the increasing regularity of these events.

I have never seen the *acqua alta* (and have never packed Wellingtons on the off chance either), but this time I am hopeful. I have chosen the date of my trip to coincide with the autumn's highest spring tide. My plane arrives in the midst of a savage thunderstorm, but the light of the full moon breaks through the clouds when we land. By the next morning, more than an inch of rain has swollen the waters of the lagoon, increasing my expectation still further. I make for the Riva degli Schiavone, the broad embankment that looks out onto the island church of San Giorgio Maggiore and the lagoon beyond. The water rises and laps tantalizingly at the marble curbs. A few waves kicked up by the passing vaporetti do splash over, but the sea remains a few millimeters below the edge. I notice that the path, in fact, slopes down away from this edge, so that if the water were to rise any further, it would quickly flood both the path and the restaurants behind, and begin its inexorable dash through the streets and into doorways. Studying the quayside stonework more closely, I see that it has already been raised more than once against the sea.

The Venice barriers present a greater technical challenge than the Thames Barrier or even the Oosterscheldekering for two reasons. They must allow the passage of ships eight times larger than those that use the Thames. In addition, because they will stop the three mouths of a large, shallow lagoon that relies almost entirely on seawater flushing rather than river effluence to maintain its depth and life—and because of increased scientific and public awareness of the environment since those earlier schemes were conceived—the Venice barriers will have to show unprecedented sensitivity to the local ecology. The Oosterscheldekering, for example, has recently been found to act as a trap for porpoises that venture into the sea behind its open gates. "They enter it, but cannot or apparently do not dare to swim out again," explains Okka Jansen at Wageningen University, who has studied the mammals' movements by analyzing their stomach contents. The ideal objective is that the Venice barriers should offer the minimum obstruction to the tidal flow when they are not being used, but complete security against the sea when they are. As the leading tidal oceanographer Walter Munk put it in the title of an influential paper, the idea is to "Let the Moon Sweep the Lagoon"—as it always has done.

Predictably, the massive project has lately become the focus of a corruption scandal, with many senior consortium executives and government officials—even including the mayor of Venice—placed under arrest. However, construction itself is well under way. The barriers are now expected to go into operation in 2018, more than twenty years later than first hoped. The week before I arrive, a first test of four of the barrier units in situ—a full-scale version of the models I had long ago seen outside Rome—has apparently gone well, although the concierge at my hotel, who, like everyone in this city, has family involved in the project, gives his version of the story with much shoulder shrugging and eye rolling.

In the offices of Consorzio Venezia Nuova this time, behind the high brick walls of the Arsenale, I meet Giovanni Cecconi, who will have responsibility for operation of the barriers. He has kindly agreed to see me, even though he is just off the red-eye from New York. He has found a crisp, white shirt and is wearing the rimless glasses of an engineer. He is keen to show me the presentation he gave in New York, but the computer won't cooperate, and instead he distracts me with the pictures of his family pinned to the wall, whom he perhaps hasn't yet had time to see.

There is a change of emphasis since my visit twenty-five years ago. It is clear now that preserving the lagoon and saving the city are regarded as one and the same thing. In 1992, progress on the engineering side of the scheme was held back when a national law was passed demanding a greater emphasis on environmental restoration. Fabrication of the barriers finally began in 2003. As Giovanni tells me, the fortunes of one depend on the other, as they always have. It is the combination of "protectedness and openness" of the lagoon that has always given Venice its unique character. It is the treacherously shallow lagoon that allowed Venice to resist amphibious advances by other states. (The Venetians simply pulled up the posts that mark the safe channels at times of threat.) But it is the lagoon as giant harbor that also welcomed foreign trade and cultures, each of which added to the "flavors" in the city. "Any time you see something beautiful in Venice, you see this combination of protectedness and openness," Giovanni says proudly.

Giovanni goes on to tell me about the creation of barrier islands using the natural forces of tide, wind, and waves to allow them to form where they will. In the right places, they will trigger "biostabilization"; seaweeds will come, and then grasses, which will trap dust and enable the islands to grow. These islands will not necessarily be where they once were, because the lagoon is different from the Middle Ages when Venice rose to greatness and the rivers flowing into it were first diverted for freshwater. "It's impossible to go back; it's evolving," says Giovanni. The object is to recover the natural processes of the lagoon, not its former shape.

And what was he doing in New York? There, he was part of a Venetian delegation (including the shortly-to-be-accused mayor) meeting New York mayor Michael Bloomberg in order to discuss that

city's resilience to attack from the sea. New York State established a
Governor's Office of Storm Recovery in the wake of the damage done
by Hurricane Sandy in October 2012. Inspired by the Venetian example, the city is looking at the feasibility of an engineering scheme that
will not simply wall up Manhattan against the sea, but will improve
the resilience to storm surges of the whole area, including the Upper
and Lower New York Bay and the Rockaway coast. It is a reminder
that New York, like many coastal cities around the world, is a potential Venice.

I LEAVE and catch the vaporetto to re-join the tourist crowds. I still
want to know more about the barriers that Giovanni spoke so little
about. I need to see concrete and steel to convince myself that Venice
will be made safe, and for this I must make a longer journey across
the lagoon. We speed past desolate brick fortifications, neglected
islands with scrubby little trees, and mud banks with reddening samphire, our wash the only thing disturbing the glassy gray water. In
spirit at least, the lagoon can have hardly changed since Galileo's
time. Fishing goes on as it has done for centuries, with only the skin-
niest of poles to indicate the presence of nets strung *aseriagi e con covoli*
("in series with intermittent traps") in water that is only 2 or 3 meters
deep. No wonder the channel is so well marked, with its clusters of
four posts leaning conspiratorially in to one another every 100 meters
or so, generously topped with navigation lights and radar reflectors.
Here and there, in the middle of the burnished expanse of water, is
a lashed-up shelter of timber and reeds erected to offer fishermen in
their boats protection from the sun.

At last we reach the sleepy town of Chioggia near the southernmost of the three entrances to the lagoon. It is Venice without the
tourists, Venice as it once was. On the peninsula leading out to the
lagoon entrance, the road reaches a dead end at a beach café. From
here, I walk out along a newly built breakwater that lines the south

side of the entrance channel. The marble path is flanked by break-waters made of gleaming white rocks each the size of a Fiat 500. In the distance are the elaborate rigs where the fishermen dry their nets. They make the place look already traditional. Across on the north bank of the channel, out of the reach of the sightseers, is the island building site where the gargantuan components of the barrier are being prepared. I can see two monolithic blocks of concrete each as big as a hotel—one under construction with scaffolding around it, the other perhaps ready to go. These are two of the bases for the hinging segments of the barrier. When the time comes, the massive dugout basin in which they have been constructed will be flooded, and they will be carefully floated out into the channel and deposited in the correct position. Yellow buoys on the water already mark where they will go.

The Venice barriers may not be "iconic" like the Thames Barrier, but neither will they be obtrusive like other schemes. These huge concrete foundations will, of course, lie unseen on the seabed; and so, for most of the time, will the seventy-eight steel elements that make up the hinging barrier gates. Even the control buildings have been designed not to rise above the level of the land. It is an audacious and stylish statement of confidence in a project whose purpose is to defeat a natural disaster for which human instinct is to escape to the heights.

AS I PREPARE to leave Venice, I notice again the still-full moon. How could Galileo have left it out of his theory? Surely the coincidence of the phase of the moon and the range of the tide observed by every primitive culture with access to a seacoast was too great to ignore. He must have believed tenaciously in the power of his own reasoning to ignore all that, and in fact he got more right than is often acknowledged. But without an understanding of the force of gravity, he had no chance of being substantially correct.

On a final stroll around town, I see that the water barges still come, only now they are unloading crates of San Pellegrino for the restaurants along the Riva degli Schiavone.

6

THAMES MUD

In 1924, the Anglo-French controversialist, onetime member of Parliament, and author of cautionary tales for children Hilaire Belloc sailed alone around the British coast. He wrote about his voyage in *The Cruise of the Nona*, which Jonathan Raban has described as "the weirdest possible blend of Mein Kampf and Yachting Monthly." Long misanthropic rants—against atheists, pacifists, Jews, all who lack "virility"—and unbidden paeans to Mussolini are interspersed with short but vivid descriptive episodes as Belloc and his boat wrestle with the waves.

Somewhere off the coast of Devon, Belloc is moved to fulminate against the scientists who would know the oceans. "No man living can understand the tides. And the mystery of the tides is as good a corrective as one could find to our deadening pride in exact measurements, and to the folly of attempting to base real knowledge upon mere calculation: our pretence to a universal science, and to a modern omniscience upon the Nature of Things." He goes on to offer his sarcastic surmise of the "bungling of landsmen in the matter of tides" (he means men such as Galileo and Newton), pointing to various "anomalies" that he has experienced on his journey—double tides, long highs, swift floods, sudden ebbs—that seem to him to obey no rule but God's. "All this questioning sounds like the Book of Job; but, note you, that I, for my part, am with Job, and against the scientists."

Except that, of course, he wasn't. Not really. For it is clear elsewhere in the account of his cruise that Belloc, like any conscientious coastal sailor, consulted the published tide tables, and surely knew that his life largely depended on them.

I AM BACK ON THE BANKS of the River Thames at Greenwich, a little way upstream from the Thames Barrier and downstream from London Bridge, where England's earliest tide table was made. This riverside walk, I see from a map released by the Environment Agency in the aftermath of the severe North Sea storm surge in December 2013, would have been underwater along with large parts of East and South London if it were not for the barrier.

Watching the brown water slide past the moored barges, I am curious as to how London became a major port. It is inconveniently far from the open sea; in the days of sail, it would have been almost impossible for a ship to leave its mooring and get all the way downriver using the assisting current of the ebb tide—more than "a tide's work," in an old sailor's phrase. Today, the *Cruising Association Handbook* still advises yachtsmen, "Even starting from Limehouse with no more than average speed, a foul tide will be met long before comfortable shelter can be found outside the Thames." With the prevailing wind in the west, sailing upriver would take even longer, perhaps a day or two, riding out the foul tides at anchor, and necessitating the use of more than one flood tide. Tidal movement is essential to prevent stagnation of a river long used as a sewer, especially during dry summers when little freshwater comes downstream to augment the ebb. To humans it is an advantage too, but it is one that must be seized and ridden quickly, before the direction changes. No wonder the first Anglo-Saxon landings stopped some 20 miles downstream. The Dutch fleet that raided in 1667 contented itself with capturing and destroying fifteen naval ships from the dockyard at Chatham, and felt no obligation to wait for the next flood to go further up toward London to make its point—yet another occasion when Britain was bested by its own tides, to add to the Battles of Fulford and Maldon.

This is Charles Dickens's stretch of river too. He was born by one sea—in Portsmouth—and died at his home in Kent, not far from where the estuary of the Thames broadens out to meet another. He understood the tides like few other writers, even those given to sea-faring tales, and made them integral to the atmosphere and the plots of his novels. The River Thames is a forceful presence as both a high-way and a clock. It is also a narrative convenience, its tides delivering some characters and providing the means of disposal of others. Even those working in offices find it is the Thames that sets the rhythm of their lives. Sometimes, its ceaseless motion is a cipher for their own busyness; at other times, their torpor is reflected in the endless cycle of its ebb and flow. For a few of Dickens's characters, the river provides a dubious livelihood.

Our Mutual Friend opens on the river, "between Southwark Bridge which is of iron, and London Bridge which is of stone," with Gaffer Hexam and his daughter Lizzie recovering a body from the water for the valuables that it might have about it. The pair seize the ebb tide to spirit the body away from the city and downstream to the Surrey side. While Lizzie rows, Hexam scans the rushing water for other finds as is his habit: "At every mooring chain and rope, at every stationary boat or barge that split the current into a broad-arrow-head, at the offsets from the piers of Southwark Bridge, at the paddles of the river steamboats as they beat the filthy water, at the floating logs of timber lashed together lying off certain wharves, his shining eyes darted a hungry look."

The Thames transports bodies, dead or alive. Early on in *Great Expectations*, the orphan boy Pip imagines himself whisked off "on a strong spring-tide" to the prison hulks that lie moored in the river close to his guardian's smithy in the marshes. It is a kind of guilty premonition of a key scene at the climax of the book when Pip, now a young man, takes a rowboat in an attempt to help the criminal Magwitch escape his pursuers. Pip uses the ebb tide to carry him down-river, where he will be able to pick up his illegal passenger and then discreetly put him aboard a steamer bound for the Continent. At one point, Pip imagines that the tide—the tide of the river, the tide of events—is running toward Magwitch and that "any black mark upon its surface might be his pursuers, going swiftly, silently and surely, to take him."

It is to be a long row. Pip and two friends set off at high tide from Temple Stairs in the middle of London and are carried swiftly down-stream through London Bridge—at that time still the river's lowest crossing. (Pip has previously practiced his oarsmanship to be able to negotiate tricky passages such as the tidal race running between the piers of the bridge.) They have six hours of the ebb to help them. They pick up Magwitch and another convict and then row on—against the tide now, which has begun to flood—before putting up for the night in an isolated pub to await the Hamburg steamer due the next day. On the following morning's high tide, they row out to meet the steamer but are apprehended by a police galley. The two small boats lock together and swing around in the tide into the path of the fast-approaching ship. There is a brief moment of confusion as Pip and his accomplices are taken aboard the police boat. But where is Magwitch? All scan the water. "Presently a dark object was seen in it, bearing towards us on the tide." As Pip's river nightmare foretold, it is Magwitch, injured—fatally, as it will turn out, as he tried to swim to safety—by the churning paddle wheels of the steamer.

And in *David Copperfield*, when the Yarmouth fisherman Daniel Peggotty sees that Barkis, the carter who married his sister, is dying, he repeats the folklore long ago recorded by Aristotle and Pliny: "People can't die along the coast . . . except when the tide's pretty nigh out. They can't be born, unless it's pretty nigh in—not properly born, till flood. He's a going out with the tide."

NEWTON'S *QUAESTIONES*

Up the hill from the grand symmetries of Christopher Wren's white stone naval college and hospital is an older building of his, the Royal Greenwich Observatory, completed with the assistance of Robert Hooke more than twenty years earlier in a very different style—a pleasingly domestic jumble of red brick.

Even before Greenwich gained its formal association with time— the prime meridian by which the world still sets its clocks was globally adopted in 1884 and is scored into a brass plaque here—it was a place greatly concerned with the tide. The house on the hill was built for the first Astronomer Royal, John Flamsteed, appointed by Charles II in 1675. It was laid out according to Flamsteed's instructions to suit his life of science. His life's work—achieved despite chronic headaches and a meager budget—was the assembling of a star catalogue, the posthumously published *Historia coelestis*, which listed the positions of over three thousand stars, based on observations he had made during a period of more than thirty years using telescopes that, despite his illustrious job title, he was obliged to acquire at his own expense. These data were a valuable contribution toward solving the so-called longitude problem that would eventually enable His Majesty's ships to know their positions at sea.

But this was nighttime employment. Flamsteed doubtless had frequent occasion to journey between his hilltop observatory and meetings in London, for which a riverboat would be the natural choice. It is perhaps no great surprise, then, to find that soon after his installation at Greenwich, Flamsteed, along with various "ingenious friends," also began making direct observations of the Thames tides. The crude approximation—seen in the old St. Albans Abbey tide table and still repeated by the seventeenth-century almanacs—that high tide fell three hours after the moon passed through north was no longer good enough. It worked well at spring tides (when the moon was full or new) and neap tides (when there was a half moon) but was much less accurate for the tides in between, when the moon was in its crescent or gibbous phases. From 1683, therefore, Flamsteed began to publish his own improved annual tide tables. In an effort to unite his empirical observations with astronomical theory, he also made more precise measurements of the orbit of the moon, measurements that were soon to be coveted by a man called Newton.

Isaac Newton was a landlubber. It seems that never in his long and illustrious life did he leave England's shore. (Flamsteed at least made a voyage to Ireland as a young man.) It is quite likely that the only tides Newton had physical experience of were the ones he too occasionally had reason to ride up and down the Thames.

As theory, though, the tides interested Newton intensely. In 1664, while studying at Cambridge, he began to list the *quaestiones* that would dominate his life of unparalleled scientific inquiry. One of them was to explain the tides. Aside from Galileo's oddly moonless theory, the most influential new thinking about the cause of the tides at this time came from René Descartes's theory of vortices, which attempted to explain the tides as an effect of the pressure exerted on the earth's thin ocean layer by the moon. Newton's thought was to test Descartes's theory by comparing the tides with air pressure as measured by barometers. If there was a correlation, this would support the French philosopher's theory; if not, well, something else must be going on. The primary task was this: "Observe if ye sea water rise not in days & fall at nights by reason of ye earth pressing from [moon symbol] uppon ye night water &c." Newton also sought to establish whether the tides were higher in the mornings or evenings, or at particular seasons, and how all of this related to the earth's orbital velocity and distance from the sun.

Of course, Newton never undertook these exercises directly. But they were the thought experiments that enabled him to reason out his theory of gravitation using mathematics and pure reason. The answers to many of his *quaestiones* were eventually synthesized in the *Principia*, published from 1687 onward in three volumes setting out the fundamental laws of motion and a comprehensive theory of gravity, which together explained the observed motions of celestial bodies far more satisfactorily than ever before. Book I of the *Principia*, *De Motu corporum* ("On the Motion of Bodies"), was mainly mathematical; Book II was mainly text, discussing the application of these universal rules to the sun, earth, and moon, and the rest of the solar system. Newton added the third book as an afterthought in order to provide a more rigorous development of the relatively accessible discussion in Book II and as a means of silencing his critics, not least among them his enemy Robert Hooke, whose cursory rebuttal of his ideas about the nature of light years before had led to a bitter feud between the two men. Newton's theory of gravity drew some early criticism: how could there be a force of attraction between bodies unless there was some medium through which the force passed, as in Descartes's theory of vortices operating through the ether? But its accuracy and completeness ensured that it ultimately swept all before it.

Much of the analysis in the *Principia* concerned the motion not of unimpeded bodies in space, but of bodies, including fluid bodies, that experience resistance. Newton's case studies included the speed of travel of waves and the way water flows out of a cylindrical vessel with a hole in the bottom. The tides were a practical extension of these abstract problems. Unlike these problems, however, they could not be explained by reference solely to local forces. A proper analysis also required a solution to the three-body problem of the earth, moon, and sun treated as one mutually interacting system. The masses and relative distances between these bodies had been calculated previously. Now, Newton's equations of motion and gravity employed these quantities to show for the first time how the earth's tides vary according to the position of the sun as well as the moon. Newton was able to quantify the relative separate contributions to the tides of the two bodies, concluding that the massive sun exerts nearly half as much tide-generating force on the earth as the much closer moon.

In the case of tides, the theoretical science was taking shape in

Newton's head almost faster than it could be written down. No sooner had his copyist, one Humphrey Newton, a distant relative, scripted a paragraph, than Isaac would strike it out and replace it with a more detailed mathematical discussion ten times the length.

But he also needed data. For example, Newton established the ratio of the tidal forces due to the sun and the moon from measurements of spring and neap tides made at Bristol and Plymouth—one of the first data-gathering projects set up by the new Royal Society, published in an early issue of its *Philosophical Transactions* in 1667. In the mid-1680s, when he was working on the *Principia*, Newton wrote repeatedly to Flamsteed, pestering him for his latest observations of comets and planets, to see whether the astronomer's data would conform to his emerging theory. He wrote again when Flamsteed published a new table of tidal observations in the *Philosophical Transactions*. Flamsteed acceded to these demands, though he was often baffled as to how his data could be of any use to Newton. However, Newton was now able to expand his treatment of tides in the *Principia* with reference to Flamsteed's data and so could go into greater detail concerning various perturbations and inequalities in the motion of the moon, which would have their own implications for tidal calculations. But he perhaps omitted to give Flamsteed his due, and the Astronomer Royal was later to regret having been so free with his data.

The *Principia*, however, was a triumph. Even those who did not understand it—and there were many—hailed it as a masterpiece. The world of physics was changed forever; indeed, the world itself was seen anew. Not a few picked up the book because they wanted to read an explanation of the tides above all. One of these was Gilbert Clerke, a noted mathematician and philosopher sometime earlier retired from academic life in Cambridge, who found it hard-going. As a former mariner, Edmond Halley, the astronomer who had guided and sponsored its publication, was also acutely interested in Newton's revelations concerning the tides. He, in turn, sent one of the first copies of the *Principia* to the new king, James II, who also had maritime interests, having previously been Lord High Admiral of the Royal Navy. Halley included a cover letter drawing the monarch's attention especially to Newton's tidal analysis.

THE COSMIC LEASH

It is almost impossible to overstate the importance of Newton's break-through. The economist and Newton enthusiast John Maynard Keynes wrote on the tercentenary of the physicist's birth, "Isaac Newton, a posthumous child born with no father on Christmas Day, 1642, was the last wonder-child to whom the Magi could do sincere and appro-priate homage." Keynes is only a little more extreme than many others who raised Newton to the status of a demigod of rationalism. Newton earns this praise deservedly for his revelations of the nature of light, his explanation of the laws of mechanics, and his development of dif-ferential and integral calculus in mathematics.

But the law of gravity seems all the more remarkable for being hidden in plain sight. It affects us all the time and is yet something "in which the common people do not even suspect a mystery," said Voltaire, another of Newton's illustrious champions.

In *Don Juan*, Byron writes,

When Newton saw an apple fall, he found,
In that slight startle from his contemplation—
. . .
A mode of proving that the earth turn'd round
In a most natural whirl, call'd "gravitation;"
And this is the sole mortal who could grapple,
Since Adam, with a fall, or with an apple.

In fact, Newton's first thoughts on the nature of gravity occurred while he was still a young man studying at Cambridge. He was able to mull over the problem at his family home in Lincolnshire, where he retreated when the plague came to the university in 1665, and where the garden claims even now to have the fabled apple tree. But it required twenty more years of scientific graft to give the theory its full mathematical development, during which time Newton was able to refine his ideas, taking advantage of new experimental evidence and observations—of comets and planets and much else—being gathered by others. This support enabled Newton to demonstrate the universal application of the law of gravity not only to falling objects on earth and the tides, but to the solar system beyond.

And how weird this domestic, cosmic force is. The resistance Newton met with when he finally published his theory of gravity is surely understandable: he had proposed that bodies here exert an influence on bodies there—mere feet or millions of miles away—without any medium whatsoever to transmit that force between them. How could a force make itself felt through *nothingness*?

Gravity seems weirdest when you stop to think about it, as Newton did. Most of the time, it seems quite natural. But perhaps this is only because we have, in fact, already invested so much time in trying to make sense of it. If you watch an infant grow to become a toddler and then a child, you may come to the conclusion, as I have done, that we spend a large proportion of our first months and years on this earth studying the force that keeps us on its surface. We drop this, we let fall that; we drop and laugh, drop and frown, drop and cry; we let fall, let fall again, and again and again, reacting variously with laughter, tears, and bafflement. Why does everything fall downward? Why not up, or sideways, or all ways, or not at all? In the end, it is still deeply strange that all falls downward.

Although the theory of gravity did not come to Newton in a single moment of inspiration, it was this weirdness that Newton suddenly saw again through childlike eyes, whether an apple ever truly fell from the tree in his garden or not. (Though possibly based on Newton's own anecdotes, this particular myth was not in circulation until 1726, the year before he died, helped on its way by Voltaire, among others.)

A few writers have thought to grapple with this weirdness. In *Ratner's Star*, for example, the novelist Don DeLillo writes powerfully of gravity as "sheer *wanting*": "There is *want* at the center of the earth." This pull is so universal in our experience that we seldom sense it in this visceral way. When things fall to earth—apples, the balls supposedly dropped by Galileo from the leaning tower at Pisa, meteorites, a plate of buttered toast—they do so with casual abruptness. The magnitude and majesty of gravitational force would be better conveyed if we could see something falling *up*, and the only body that we can regularly witness doing this is the mass of water of the oceans moved by the moon's pull: the tide. Then, as the Suffolk writer Ronald Blythe describes it in *The Time by the Sea*, we behold "the sea on a cosmic leash."

THE MATHEMATICS of the earth's tides that Newton elucidated may be intimidating, but the principles are simple enough, even if they are highly counterintuitive. "In some senses," the MIT physicist Richard Lindzen tells me, "the tides are the easiest problem in geophysics. It's not often that you know the major variables with such certainty."

You may be relieved to know that I will leave the mathematics aside here, noting only that, for some, it is the very susceptibility of such a pervasive natural phenomenon to mathematical treatment that provides the incentive for studying the tides in the first place. Kevin Horsburgh, for example, head of marine physics at the UK National Oceanography Centre in Liverpool, has always loved the sea. He has dived, and sailed, and always tried to ensure that he is never far from the coast. But above all, it is the sea as a potentially soluble problem that grips him: "I love the fact that such a socially significant phenomenon submits itself so readily to maths." More than once I had been told that those oceanography graduates who unaccountably lose the conscience about the natural environment that once led them to study the subject at university often find jobs in investment banking, where their skill at juggling dozens of interdependent variables finds a new relevance and, of course, greater financial reward.

We can avoid mathematics, but we will need to exercise some visual imagination. Newton's concepts of gravitation and planetary motions explained what Galileo's theory did not—why, in most places, there are generally two tides a day. If we could view our planet from a distance, we would see that the oceans are forced into a bulge not only on the side facing the moon, pulled up by its force of attraction, but also on the side directly opposite (although, in fact, the scale of the tides is so small—a matter of a few meters compared to the earth's diameter of more than 12,000 kilometers—that the bulges would be invisible to the naked eye, and in any case this neat theoretical picture is compromised by the presence of continents breaking up the oceans covering the planet). Because the earth spins on its axis every twenty-four hours, this double bulge is what we experience as the tide roughly every twelve hours.

"The two opposing bulges in Newton's simplified model is the hardest thing to convey to any audience," Kevin warned me in our interview. Hilaire Belloc goes so far as to accuse Newton of cooking up the formula to explain it, which he condemns as "false dogma," saying that if the tides were daily instead of twice daily, Newton would have fiddled his figures "to explain *that* with just as much ease." And indeed, it does seem odd that, if we start with an ideal model of a spherical globe covered with seawater to a uniform depth—which is inaccurate in many ways, but is nonetheless a useful simplification of the kind that physicists must often make at the beginning of their journey to understand something vastly more complex—and then put a moon in orbit around it, we will end up with a planet that is not egg-shaped, with one bulge, but the shape of a rugby ball, with two.

To see why this is so, it is helpful to take a step backward. Imagine this same ideal earth, with the moon brought up to an appropriate distance from it. Both bodies are stationary for now—not spinning on their axes, and not orbiting one another. The moon will exert a gravitational attraction on the earth, and the earth on it. The moon will also exert a gravitational pull on the water wrapping this earth, causing it to flow toward the side facing the moon, and raising there a permanent high tide. This force does indeed cause the waters to bulge only on the side facing the moon; the earth is like the yolk, and the oceans are like the white of an egg. However, this model is not so much ideal as impossible. For in this stationary reality there is, of course, nothing to prevent the earth and moon pulling together until they collide.

Now we can begin to introduce some of the complications of reality, one by one. We'll stop well short of the end of the list too, since many minor factors must be taken into consideration for a full explanation of the tides. "In ALL of the motions of these orbs there is not, to my knowledge even one perfect circle; not one uniform speed of rotation, not one wobble-free axis of rotation. The modern clock is much more uniform than any earth, moon, sun occurrence. Further, there are not two relative positions of earth or moon that recur with clockwork precision. . . . All the movements are predictable but none make a simple rhythm," warns the author of my field guide to the New England tide marsh, Mervin Roberts.

The most important thing to do next is to put the moon into orbit around this ideal earth. This orbit is what maintains the distance between the earth and moon. In fact, because gravitational attraction is mutual, the earth also orbits around the moon; or, more accurately, both bodies orbit around their mutual center of gravity, which, given the respective masses of the earth and moon, actually lies inside the volume of the earth, some two-thirds of the way from the center toward the surface. In this new situation, tidal forces raise a tide on the side of the earth facing the moon as before, but there is now also a new force of acceleration to be taken into account, acting on the earth in the opposite direction, away from the moon, owing to its orbital motion. Because of its own inertia, the water on the surface of the earth tends to be left behind by this constant acceleration. In Newton's simple model, this lag raises a new bulge on the side of the earth facing away from the moon. Newton's analysis not only showed how this was the case in qualitative terms, but also proved mathematically that the two tidal bulges are equal in magnitude.

We now have an earth with two regions where there is a permanent high tide and a circumferential band in between where there is a permanent low tide. Now let's set it spinning on its axis. For the sake of simplicity, again, let us say that this axis runs at right angles to the line between the earth and the moon. (We know, of course, that in reality it is tilted.) The earth now whizzes around beneath its covering sheath of water, which remains essentially static above (we are assuming there is no friction between the water and the earth). Now, the picture is that every place on the earth experiences two tides a day. These tides are greatest at the equator and taper away to nothing at the poles.

WATCHING THE TIDE on the Norfolk coast, I wondered what happens to a given cubic centimeter of water. How does it move? Newton turned this idle speculation into a powerful thought experiment. In the *Principia*, he begins, as we have seen, by imagining the force on a solid body in orbit around another solid body. In order to address himself to the tides, he exchanges this solid body for a particle of liquid. Then, he imagines lots of these liquid particles that join up to form a ring of water. Next, he imagines the second body, at the center of this ring, expanded until its surface touches the water, and finds that the water is accelerated along the ring. (Newton's diagrams are two-dimensional, but the three-dimensional extension is a water-covered sphere.)

Newton did not stop there, however. He made arguably his greatest contribution to our understanding of the tides as they are actually experienced by demonstrating for the first time how the sun contributes to the tides, thereby explaining the difference between the high spring tides that occur at full moon and new moon and the lower neap tides around half moon. Again, the principle is simple enough. When the sun and moon are close to alignment—either both on the same side of the earth (new moon), or nearly in line on opposite sides (full moon)—the gravitational attraction owing to the sun is added to that of the moon and increases the tides. When the sun is at a large angle to the moon, they pull in different directions, detracting from the lunar tide and reducing the heights attained.

His mathematical approach even enabled Newton to put a figure on the relative influence on the tides of the sun compared to the moon. Although the force of gravity famously operates according to an inverse-square law—in other words, the attractive force is multiplied fourfold when two bodies are placed half as far apart—the force that raises the tide is inversely proportional to the *cube* of the distance. If an inverse-square law applied, the solar component of the earth's tides would be far greater than the moon's, and we would have to reckon with far greater forces at sea. As it is, the sun's contribution to the tides is just less than half that made by the moon.

In addition, because the earth's orbit around the sun and the moon's around the earth are not perfectly circular, but are in fact elliptical, Newton was able to explain why spring tides are often a little higher and neap tides a little lower in the Northern Hemisphere winter, when the earth is relatively closer to the sun. He also showed why in summer the daytime high tide is usually greater than the one at night because of the higher angle of the sun and moon in the sky, and in winter it is the other way around. On my day observing the tide on the Norfolk coast, I stayed long enough to see the evening tide rise to roughly equal the morning's high water. (I was there in September, when there is no net effect.) I had never appreciated this systematic variation, however, through many summer and winter visits to the shore. I check now in my annual Norfolk tide tables and find that Newton's logic is indeed confirmed in the prediction figures.

Newton was able to take into account various other real-world complications as well, but he was not correct about every detail of what causes the tides, and he had to leave much else out of his reckoning. Worst of all, his theory, though a great advance, still offered no substantially improved basis for the practical prediction of tides in real harbors, at Greenwich or many other places.

THE HUNT FOR LUNAR DATA

LONDON BRIDGE

	LW	HW	LW	HW
September 1, 1694	00:09	06:30	12:42	19:06
	1.4 m	5.7 m	1.6 m	5.7 m

The learned men of the Royal Society felt duty bound to keep niggling away at the problem in whatever way they could. In his later years, Edmond Halley went back to sea—he had earlier undertaken an epic voyage to St. Helena to study the stars of the Southern Hemisphere—and made a survey of the English Channel, observing tide times at sea as well as close to the coast and measuring the tidal currents. (These investigations also encouraged his own early conjecture as to the tidal picture when Julius Caesar attempted his invasion of Britain.)

For his part, Newton continued to refine his lunar theory long after the *Principia* was published, returning to the subject intermittently through the 1690s and for much of the rest of his life, in order to incorporate various perturbations of the moon into the already complicated three-body problem. The inadequacy of his mathematics for dealing with real tides must have been rammed home, if not to him, then certainly to the navy and other sailors, by some severe storms. In October 1697 the North Sea produced its highest storm surge of the century, and then on the night of December 7–8, 1703, perhaps "the severest storm of which we have any good account" devastated southern England. Two days later the wind turned to the northwest, and an exceptional surge of water rose up the Thames. None of these events could have been predicted to any useful extent, even by the best science of the day.

Prompted by challenges from rival scholars, Newton introduced mathematical lemmas, or subsidiary propositions, to deal with awkward facts, such as that the earth is not uniform in density and is not perfectly spherical, which alters its gravitational effect. Throughout, Newton was obliged to use actual tidal and lunar measurements in order to tweak his theoretical conception of celestial motions. One glaring discrepancy, for example, was the fact—long known to those living by the coast—that the highest tides occur not precisely when the sun and moon lie most closely in a straight line with the earth as theory predicted, but a day or two afterward.

The best lunar observations were those made by John Flamsteed at the Greenwich observatory. Newton had already annoyed him by pestering him for advance data that he might put in the first edition of the *Principia*. Now, he felt compelled to renew their tetchy dialogue in what was to blow up into one of the most vituperative and bitter disputes in all science. Newton was by now in his seventies, and his always cantankerous attitude had mutated into despotic viciousness as he rose to positions of power in the scientific and national establishment. Flamsteed was still at work on his star catalogue, increasingly aware that there was as yet little to show for his years as Astronomer Royal, but convinced of the value of his project and anxious to see his life's work properly rewarded.

On September 1, 1694, Newton paid a visit to the Greenwich observatory, on the hunt for lunar data. Flamsteed obligingly showed him some figures and promised there was more to come. Somehow, Newton persuaded the astronomer to make more lunar observations of the kind he needed, even though it would take time away from the star catalogue. In December 1698, Newton chased Flamsteed for the information. Relations between the two men had not improved in the intervening years. Inadvertent errors in Flamsteed's data had enraged Newton, and the meeting merely fueled both men's paranoia, with Newton telling associates afterward that he would not be able to complete his theory, because of Flamsteed.

On April 12, 1704, Newton, now the president of the Royal Society, went again to Greenwich, blaming Flamsteed and his observations for the still inadequate lunar theory. Flamsteed at this time was in search of royal support for the publication of his star catalogue through the offices of the Royal Society. With his own intense interest in the project, Newton used the power of his position to skim off the data he wanted—still far from complete, so far as Flamsteed was concerned—and pushed ahead publication of Flamsteed's catalogue in a form that its author did not approve. Newton put himself in the role of proof-reader, making his own "corrections" to Flamsteed's readings. Halley, another enemy of Flamsteed, added amendments of his own.

Stung into action by this appalling treatment, and aware that he was nearing the end of his life, the Astronomer Royal took belated steps to see that the *Historia coelestis* would finally be published in full in the way that he had always envisioned. He also found an opportunity to gather up the undistributed copies of the unauthorized version, which he promptly burned, making "a Sacrifice of them to Heavenly Truth." But Newton had the bitter last laugh in this sorry saga. In 1713, he published the second edition of *Principia*, removing references to Flamsteed wherever he could.

MUDLARKING

On the foreshore in front of the naval buildings at Greenwich, there are signs of the past between the tide lines. They have not been carefully preserved as "heritage"; they are simply ancient items once discarded and now discovered as the river has redistributed its sediment. It is possible to explore this sometime land simply by hopping over a railing and taking the seaweed-covered steps down from the embankment. But I want to learn a bit about what I see and have arranged to join the Foreshore Recording and Observation Group (FROG) organized by the Museum of London.

My first thought had been to sample the life of the mud lark—the near-destitute persons who lived by selling what they could find thrown overboard onto the mud from moored ships when London was the world's greatest port in the eighteenth and nineteenth centuries. I have arrived here by walking along suitably picturesque Victorian back streets. Indeed, around one corner I find my path blocked by a film crew making a new television adaptation of Kate Summerscale's *Suspicions of Mr. Whicher*. A man with a paddle is pumping artificial fog along the cobbled lane. I feel right in period.

But I have a feeling that today's activity will be of an altogether higher class.

Before Wren's naval college was built, this was the site of one of Henry VIII's palaces. Part of the foreshore contains the posts of a Tudor jetty and is a designated ancient monument, off-limits to our amateur fingers. But from anywhere else, we are free to pick up what we find. "The foreshore is eroding, so we don't need to do any digging," Helen, the FROG coordinator, explains. "Unusual for archaeologists!"

There is abundant evidence of the past life of the riverside. Some large flagstones indicate the site of an old causeway, which was uncovered only when mud was washed away, partly by winter storms, but partly, it is thought, by the wash from superfast river cruisers. There are also the remnants of massive blocks of chalk lying palely on the beach that were once used to provide hardstand where barges could be run aground for unloading before the next high tide. More surprising is the sheer quantity of waste material that has been dumped here over the centuries and hasn't been dispersed by the currents or buried in mud: oyster shells, fragments of clay pipes, and large stretches where the beach is made up almost entirely of butchered beef bones. Helen points out some less obvious features, such as roots of trees, the remains of a prehistoric forest that stood on this site before human settlement.

After a while, scanning the mud and pebbles, I find a Tudor pin. They are apparently very common. It is a rather plain specimen, one of a type used to hold together unsewn cloths as garments. I look more closely at the design and see that it is clearly of preindustrialized manufacture. The head is a small piece of turned yellow metal, knurled with one equatorial crease so that it has the overall shape of a miniature hamburger. The pin itself looks ordinary, but I find when I run my fingernails down its length that it is finely ridged, so as to better grip cloth—a nice, practical detail that is absent from machine-made pins today. I am delighted with my find, and push it through a page of my notebook for safekeeping.

In his celebrated exposé of urban deprivation, *London Labour and the London Poor*, the Victorian journalist Henry Mayhew identified "The Mud-Larks" or "river-finders" as "certainly about the most deplorable in their appearance of any I have met with in the course of my enquiries." They were made up of the very young, often orphans, but also the unemployed elderly. They made their living by crawling in among the moored barges, scavenging pieces of coal and wood for fuel and refuse that they might sell, typically gathering up what they found in an old basket or tin kettle.

> The mud-larks generally live in some court or alley in the neighbourhood of the river, and, as the tide recedes, crowds of boys and little girls, some old men, and many old women, may be observed loitering about the various stairs, watching eagerly for the opportunity to commence their labours.

When the tide is sufficiently low they scatter themselves
along the shore, separating from each other, and soon disappear among the craft lying about in every direction.

Most prized were the copper nails used to repair ships, but they were
seldom picked up, because the mud larks were shooed away from
where the hulls were being sheathed. More usual finds were bones,
pieces of iron, and rope they could sell to rag shops. Coals fetched
a penny a pot. Occasionally, a dropped tool would be found, which
could be exchanged with the seamen for food. A mud lark might make
threepence a day in this way. It was no easy life. One boy, imprisoned
for stealing from an empty coal barge, told Mayhew that he enjoyed
his week in the house of correction more. In 1848, when Mayhew was
conducting his research, he estimated there were 280 mud larks at
the most popular location by King James's Stairs at Wapping Wall.
At Greenwich and other places downstream where there were more
shipyards, he counted 550.

One of the main hazards faced by the barefoot children was broken glass. Mayhew tells of one fourteen-year-old boy who was cut in
this way and, after going home to dress his wounds, returned directly
to work—"for should the tide come up," the boy told Mayhew, "without my having found something, why I must starve till next low tide."
Another boy called on Mayhew in order to tell his story. His father
had died, and though he was able to continue attending school for a
time, supported by his mother, he had been obliged to take up mudlarking to provide for them both after she was taken ill. Mayhew
intercedes and finds the boy some employment, in which he pros-
pers and is able to rescue his family's situation. Mayhew concludes,
"This simple story requires no comments, and is narrated here in the
hope that it may teach many to know how often the poor boys reared
in the gutter are thieves, merely because society forbids them being
honest lads."

. . .

ONCE AGAIN, I am walking on land that is not always there. The tidal range in the Thames is greater than anywhere on Britain's east coast, up to nearly 6 meters. The handsome seawall of the naval campus buttresses the upper part of this range, but the lower part of the twice-daily rise and fall alternately hides and reveals a broad stretch of evenly shelved beach. This foreshore is neither private, and therefore subject to laws of trespass, nor exactly public, with the sanctioned and sanitized access that this term increasingly implies. The going here is unclean, both literally and in the sense of a taboo, and my footsteps feel like a kind of transgression, although it is not clear who I am transgressing against. In ancient times, I might have been wise to offer a propitiation to the river gods. Now, I learn, it is in general the Crown Estate to which I must kowtow.

It is the state in this curious guise that owns the land between the mean high-water and mean low-water marks around about half the coast of the United Kingdom by virtue of prerogative right. This "right" is perhaps not quite as ancient as you might think; it was a ruse devised by cunning lawyers of Elizabeth I in the sixteenth century. These lines are the upper and lower legal limits of the foreshore. (In Scotland, the Crown Estate claims a slightly broader swath of the foreshore exposed between mean high and low water *at spring tides*.) The crown and state in other forms—as the Duchy of Cornwall and the Ministry of Defence, for example—has jurisdiction over much of the remainder, but charities such as the Royal Society for the Protection of Birds and the National Trust control other stretches.

This claim to ownership naturally gives rise to disputes concerning what may and what may not be taken from the foreshore. On the coast where I live in Norfolk, there has recently been a problem with theft of the foreshore itself. Some of the beaches here produce a graded shingle that it seems some people cannot resist taking for their gardens. In the past, though, it was more obvious items of value found here that were the attraction. Like the foreshore itself, these objects were subject to arcane legal classification. Today, we are familiar with the phrase "flotsam and jetsam," which is often used metaphorically to refer in a vague and all-embracing way to miscellaneous waste or surplus material, and sometimes to unwanted human populations. We are not always aware that there is a clear distinction in meaning between the two words, and that there are, in addition, other, equally precise categories to refer to matter out of place on the foreshore.

In return for keeping ships ready to go into battle on behalf of the Crown if required, the ports of Hastings, New Romney, Hythe, Dover, and Sandwich along the vulnerable Sussex and Kent coasts of England closest to France were granted certain reliefs and privileges under a royal charter issued by Henry II in 1155. In particular, as I read on a plaque by the quayside in Sandwich, these five towns, which became known as the Cinque Ports, won "exemption from tax and tallage, right of soc and sac, tol and team, blodwit and fledwit, pillory and tumbril, infangentheof and outfangentheof, mundbryce, waifs and strays, flotsam and jetsam and ligan."

Let us take a moment to savor this splendid proclamation. It is the last three of these terms that refer to the unique opportunities offered by a tidal shore. "Flotsam" describes the floating wreckage of or from a ship, whereas "jetsam" is material that has been legitimately thrown into the sea by a ship's crew—for instance, in order to save themselves in a storm. "Ligan" or "lagan," on the other hand, is cargo that has sunk to the seabed and is considered recoverable. It is to be distinguished from "derelict," which is material deemed beyond recovery and therefore does not feature in the Cinque Ports charter. The only other of these words pertinent to the topic of the tides is "mundbryce," which is the right to enter land or property for the purposes of maintaining sea defenses. Most of the other terms grant the independent right to try lawbreakers and to subject them to various unusual punishments. That other familiar phrase, "waifs and strays," refers not, as we tend to think now, to errant children, but to apparently ownerless belongings or cattle, which can be claimed after a year and a day.

Bizarrely, yet somehow magnificently too, these terms persist in modern merchant shipping law. The government Receiver of Wreck maintains a list of protected wrecks in UK tidal waters and has responsibility for regulating salvage activity. The legal definition of a wreck for these purposes includes categories of flotsam, jetsam, lagan, and derelict.

IT IS HARD TO SEE how anybody could make a living from today's riverside detritus. So much of our waste is plastic. It no longer sinks but is carried off by the tide, to be washed onto riverbanks upstream or else swept out to sea, where it is then gathered by the concerted action of currents, winds, and tides into vast and ever-growing oceanic gyres. Apart from items of potential archaeological interest—bones of uncertain age and species, clay pipe shafts, rotting timbers—I have also seen, in my short foray onto the Greenwich foreshore, colorful hanks of synthetic rope, remnants of fishing nets, numerous clouded and battered soft-drink bottles, and supermarket bags and yogurt pots.

But there are a few people who find a different kind of value in this new flotsam and jetsam. In 1999, American artist Mark Dion and a group of volunteers combed the Thames foreshore upstream of here in front of the Tate Britain and Tate Modern art galleries in central London. Over a period of two weeks, they found credit cards, mobile phones, buttons, and swizzle sticks, as well as many of the same things I see at Greenwich. The items they harvested were cleaned and presented as an old-fashioned cabinet of curiosities, making no discrimination between old and new or natural and artificial. The work, called *Tate Thames Dig*, nicely reveals the way the constant wash of the tide makes a nonsense of the stratigraphic principles upon which archaeology conventionally proceeds. The usually safe knowledge that material will be found in layers, the deeper layers being older, is obliterated. All periods are presented on the same level, thanks to the eternal tide.

Gayle Chong Kwan—half Scottish, half Mauritian, well attuned to the life of the seacoast—takes a different approach in her art. For one recent project, she has been walking stretches of the Thames from London Bridge to Margate—the riverscape of *Our Mutual Friend*—inspecting the objects that have washed up there. "People chuck these things in, knowing about the tide, thinking they'll be washed away," she explains. "But they're not. Because they're hidden at high water, they think 'invisible,' but they're there again at low tide."

Gayle leaves what she finds in situ and photographs it using a combination of large-format film and digital image processing in such a way that we lose the sense of scale that allows us to dismiss it as rubbish, and we are able to reimagine it in new worlds. The collection of images is titled *The Golden Tide*, an allusion to Noddy Boffin, the "Golden Dustman" in Dickens's novel *Our Mutual Friend*, who has grown to riches out of the waste left by others. Gayle's use of Instagram filters adds a layer of hyperreality, opening the frame of what is merely lying at her feet into broad landscape vistas that have a scope to match Dickens's moody literary description. Recognizable objects—a vacuum cleaner, a child's tricycle—become poignant wrecks. More ambiguous abstract shapes leap into unexpected life: two pieces of bright-blue plastic that have come together on the shore might be an exotic horseshoe crab. The four legs of an upturned chair in the slime become a postapocalyptic vision of London itself, like the chimneys of Battersea Power Station poking above the risen sea.

7

TIDES OF
COMMERCE

PORTS OF CONVENIENCE

On April 6, 1723, the German composer Georg Philipp Telemann heard his *Water Music* played for the first time. It was not as spectacular an occasion as the first performance of his rival Handel's *Water Music*, six years earlier, when the Hanoverian King George I and his court and musicians took barges from Whitehall up the Thames to Chelsea and back, "which his Majesty liked so well, that he caus'd it to be plaid over three times in going and returning."

Whereas Handel's piece was written to flatter the king, Telemann's orchestral suite had mercantile origins, having been commissioned, along with a secular oratorio for choir and solo singers, to mark the centenary of the Hamburg Admiralty, the port authority responsible for maintaining the channel of the River Elbe, whose tidal flushing kept the city open for business and prevented it from choking on its own sewage. And whereas Handel's was simply an occasional piece that happened to be played first on the water, Telemann's composition was a direct depiction of it.

The ten-movement suite includes sections representing various classical sea gods. But it is the penultimate movement, the gigue, that is most graphic in its depiction of the physical reality of the sea, and in particular, the movement of the waters of the Elbe. This move-

ment, which gives the whole piece its nickname of *Hamburger Ebb'
und Fluth* ("Hamburg Ebb and Flow") lasts barely a minute, but in
that time we hear a merry rippling motif on strings and oboes rise
through two octaves to represent the tide flooding in, and then fall
again just as charmingly as the tide goes out. (The conventions of
Baroque style override truth to nature here, in fact, because we hear
the rising sequence twice and then the fall twice, as music of this
period often made use of repeated passages.)

HAMBURG LIES 60 MILES or so up the River Elbe from the German Ocean, as the North Sea was then known not only to the Germans but also to British navigators. The river is broad and relatively
straight, and the land is flat, which allows the wind to blow freely
across it. All this made it practical to sail the whole long way inland
in order to unload goods nearer to major cities where better profits
might be made. These days, Hamburg is still Germany's largest port.

As in many funnel-shaped estuaries, the tidal range increases as
you go upriver, and the tidal energy becomes more focused. The difference between high and low water is about 3 meters in Cuxhaven,
which faces the open sea, but 4 meters in Hamburg itself. The difference is all the greater because there are few tributaries that might
spread the rising water and dissipate its energy. In addition, alterations to the river designed for greater ease of navigation as cargo
vessels have grown in size have made it easier for higher tides to reach
further. These changes also leave the low-lying land vulnerable to sea
flooding. On the night of February 16–17, 1962, for example, a storm
surge coincident with the high tide caused the sea to rise to nearly 4
meters above the predicted level at Cuxhaven, and 340 people were
drowned, some of them as far inland as Hamburg and Oldenburg.

In Telemann's time, when ships moved slowly under sail, the tide
was an important factor in any captain's calculation of his journey to
and from port. It was equally an important factor in preventing the

port from becoming silted up with sediment carried down the river, as well as with the waste dumped in the river by the citizens of Hamburg. Each ebb tide caught the filth before it could settle and sped it out to sea.

Listening to Telemann's celebration of Hamburg's maritime and mercantile preeminence, I wonder whether the tides were essential for the growth of the sea trade. Or were they just a nuisance that sailors learned to live with? The Mediterranean Sea was the first large body of water that became a conduit for trade; the Atlantic Ocean was the second. Two more different bodies of water can hardly be imagined— the first, generally warm and placid, with abundant small harbors; the second, stormy and violent, with adequately sheltered ports widely spaced along often treacherous exposed coasts. The contrast suggests that humankind's urge to explore, to exchange goods, and to become richer through commerce is strong enough to overcome natural obstacles of too great tides or tidelessness. It is notable that Hamburg rose to prosperity first as a member of the Hanseatic League, when it formed a trade alliance with Lübeck in 1241. Not much more than 100 miles northeast of Hamburg, Lübeck is a Baltic Sea port with negligible tides. Hamburg, on the other hand, despite its daunting tides, offers access to many more ports all around the North Sea.

In general, though, ports tend to arise where there are helpful tidal conditions, whatever those conditions might happen to be. The key requirements are not only shelter from the wind and a quiet anchorage or provision for a quay. The tides at Khambhat in northwest India, the site of the four-thousand-year-old Harappan tidal dock, have a range of 10 meters—inconveniently great for many operations, but perfect for effortlessly lifting a large ship out of the water so that its hull might be repaired. Two thousand years ago, Strabo noted the "excessive rise of tide" in the Atlantic-facing river estuaries of Spain just outside the Pillars of Hercules, observing that the tide is helpful to sailors because it scours out a longer estuary, opening up the country inland for trade. Places such as Southampton and Poole on the south coast of England grew to be major ports in medieval times because their peculiar geographical situation gave them extended periods of high tide, which made them easier to enter under sail and gave more time for unloading cargo. On the other side of the English Channel, by contrast, where the tidal range is generally greater, the first important port was Rouen, situated, like Hamburg, many miles upriver from the sea.

Furthermore, the factors that make a port attractive in one age, or compatible with a certain kind of vessel, may not be present in a later age. The general pattern of the world's tides changes only slowly, over geological time frames. But locally, there have been some significant changes. The increasing heights of high tides in the Thames and the Elbe in recent centuries have helped to prolong these cities' lives as ports. In other places, ports have silted up entirely: Dunwich in Suffolk was lost to the sea, as we have seen, although not before its river and harbor were first blocked up by storms. Near where I live in Norfolk, the walls of the Wiveton church are grooved where ships once rubbed alongside at their moorings, though the sea is now a couple of miles away.

London has ceded its position as a major port to places closer to the deep-water channels required by modern ships. Problems loom for Hamburg too. Already, the Elbe must be dredged in order to maintain access for larger ships, and this, in turn, has an effect on the tides. Such soft-edged rivers are always changing. The mouth of the Elbe has broadened owing to erosion, while riverside development— people like water views—has reduced the capacity of the banks to absorb surges. Sediment borne in on the flood tide no longer drops out into the marshes, but is carried upstream to Hamburg. In addition, the ebb tide runs out for longer because it flows more slowly than the incoming flood tide. Perhaps the changing sea levels since Telemann's day, which have caused tide heights to increase, are also responsible for this imbalance in the flows either way. (In any case, the composer gave an equal number of bars to the flood and the ebb in his gigue.) This is what hydrographers call a tidal asymmetry, and it means that with each tide, more sediment comes in than goes out. Over time, this phenomenon may cause a tidal river to be effectively lost as a human thoroughfare.

Great ports can also choke on their own success. For centuries, the tides of the Thames had been sufficient if not to remove Londoners' sewage altogether, then at least to disperse it so that it did not pose a major hazard to health. But when the city population doubled in the first half of the nineteenth century, the problem was brought home to them in the most unpleasant way. On July 7, 1855, the scientist Michael Faraday wrote in a letter to the *Times*:

> Sir, I traversed this day, by steam boat, the space between London and Hungerford Bridges, between half past one and two o'clock; it was low water and I think the tide must have been near the turn. The appearance and the smell of the water forced themselves at once upon my attention. The whole of the river was an opaque, pale brown fluid. . . . the whole river was for the time a real sewer.

Three years later, in the exceptionally hot, dry June of 1858, members of Parliament were so appalled by the "Great Stink" rising from the river directly outside the House of Commons that a law was enacted to implement a sewerage scheme for the city. The plan, developed by the engineer Joseph Bazalgette, called for underground sewers to be laid along both banks of the river to outfalls downstream at Barking and Crossness. Bazalgette's "solution" effectively shifted the whole problem downstream. Now the slick of Londoners' excrement did not oscillate between influential Westminster and wealthy Chelsea, but from the deprived East End down to Tilbury. In 1866, raw sewage helped to spread an epidemic of cholera through an area served by the East London Water Company, which had not isolated its reservoirs from the river backflow carried by the tide: on June 27, a laborer named Hedges and his wife died at Bromley-by-Bow; by August 1, 924 people had perished. Only in the 1880s did the Metropolitan Board of Works reluctantly concede that it was unacceptable to discharge raw sewage, and the waste began to be compressed into sludge to be taken by barge and dumped at sea. Bazalgette's monument on Victoria Embankment bears the discreetly evasive inscription *Flumini vincula posuit* ("He put the river in chains"), although something along the lines of *Flumine eduxit merda* would be more accurate.

THE PILGRIMS' DELAYED DEPARTURE

Not all famous sea voyages begin or end at a major port, of course. The group of English religious malcontents who came to be known as the Pilgrim Fathers famously landed in 1620 at a place in Massachusetts that they named Plymouth, although they had first briefly set foot on American soil near the site of present-day Provincetown on Cape Cod, and certainly did not disembark from the *Mayflower* by stepping genteelly out onto the famous Plymouth Rock; that was a story concocted a century later.

Although they were forced to settle many miles north of their intended landfall in Virginia, the Pilgrims' arrival in America nevertheless went rather more smoothly than their several efforts to leave the old continent, which had begun thirteen long years before. In the summer of 1607, William Brewster (a Nottinghamshire postmaster), William Bradford (a young religious radical), and a number of other nonconformists, together known as the Scrooby congregation, walked some 60 miles to Boston in Lincolnshire, where they had made arrangements to take a boat to Holland. They mustered discreetly a few miles downstream of the town on the banks of the broad tidal channel, known as the Haven, that made Boston then a major North Sea port.

Perhaps only the bleak flatness of the land here is the same today. Progressive siltation over the centuries has gradually turned treacherous marshes into rich farmland. The names of some places now well inland betray the fact that they were once closer to the sea: Fishtoft, Holbeach, Tydd Gote. It was the Romans and then the medieval barons who made the first attempts to embank and drain the fens. But it was not until the seventeenth century and the arrival of Dutch water engineers that a reliable grid of drainage channels was laid out to safeguard this land from flooding. Once, there were strict rules apportioning the length of dike each farmer must maintain in proportion to the land he cultivated. The roads across this tabletop landscape make apparently senseless zigzags because they still follow the lines of these old seawalls and dikes. I take one of these tortuous roads out of town until it leads me down to the water by an empty grass parking lot. There is little here except for shining mud, puling redshanks, the wind, and a pervading smell of cabbage. Behind the embankment is a modest stone monument. A new plaque has been recently placed on it, thanks to the efforts of an American church association. The old inscription, which I noted on a previous visit, read:

NEAR THIS PLACE

IN SEPTEMBER 1607 THOSE LATER

KNOWN AS THE PILGRIM FATHERS

MADE THEIR FIRST ATTEMPT

TO FIND RELIGIOUS FREEDOM

ACROSS THE SEAS

ERECTED 1957

In the new version, the word "made" has been replaced. Now it says they "were thwarted in" the attempt.

The Dutch vessel arrived as planned at a little creek off the main channel, but the captain had betrayed the intentions of his passengers, and as soon as they were boarded, bailiffs arrived by boat from

Boston and took the men, women, and children back up the Haven channel and locked them in Boston's guildhall to await trial. Today, the guildhall has welcome signs in Polish, Portuguese, and Russian; the display inside insists that the town treated its Pilgrim prisoners kindly, as many shared their separatist tendency.

They were soon released, and the following year, they made a second attempt to leave the country, this time from Immingham on the River Humber. The emigrants were ferried aboard a Dutch ship, but stormy weather delayed their departure, and the women and children were taken ashore to pass the night more comfortably in the local parish church. Then, word came that the authorities had got wind of the illegal escape, and the Dutch captain gave the order to set sail immediately. The tide was low, and it was impossible to board the women and children in time. Left ashore, they found themselves detained once again, but they were eventually allowed to join their menfolk in the Dutch Republic after a public outcry against the injustice.

The congregation arrived in Amsterdam and soon settled in nearby Leiden. Though they were now free to practice their religion, they found the bustling mercantile life of the city was not to their liking. Bradford was also still harried by the English authorities. Before long, they were thinking of moving on to set up their own colony in America. After raising funds for the voyage, the Pilgrims boarded a ship named the *Speedwell* at Delfshaven, and sailed for Southampton Water, where they were joined by the *Mayflower*, which had sailed from London with a passenger list mainly of artisans who would be of practical assistance in establishing a settlement. The two ships then sailed together down the English Channel, but the *Speedwell* was found to be unseaworthy, so they turned back to Plymouth, which then became the last of the Scrooby congregation's four attempted departure ports, as the overcrowded *Mayflower* now set out alone across the autumnal Atlantic Ocean with 102 passengers (including, as it happens, my ancestors John Alden and Priscilla Mullins).

All told, as many as 250 "old Bostonians" followed the first pilgrims during the 1620s and 1630s, four of them becoming governors of Massachusetts, and others, founders of Harvard University.

THE PRICE OF TEA I

BOSTON, MASSACHUSETTS

	HW	LW	HW	LW
December 16, 1773	00:27	06:28	12:41	19:05
	3.3 m	−0.3 m	3.6 m	−0.6 m
December 17, 1773	01:21	07:21	13:35	19:57
	3.2 m	−0.2 m	3.5 m	−0.5 m

A hundred and fifty years later, the Pilgrim Fathers' descendants found the tide running in their favor as they sought to throw off the shackles of British rule in America. In 1767, the British government imposed a tax on tea and other commodities imported by the American colonies. Tea was already highly fashionable by this time, and the tax was in part intended as a measure against smuggling, although this was a far greater problem in Britain itself, where tea could easily be brought in illegally from the Continent. Tea made up almost half the revenue of the East India Company, and the duty on tea raised more than 6 percent of the national budget. (Duty on cigarettes and alcohol today accounts for about 2 percent apiece.) Most of the duties were removed a few years later, but not the tax on tea. The simmering injustice set the scene for the formative colonial protest that became known as the Boston Tea Party.

The "colonials" had become adept at using the tides in the tricky waters of the New England coast in the tense years leading up to the American War of Independence. In June 1772, for example, an American packet, the *Hannah*, lured the British warship *Gaspee* into a shallow part of Narragansett Bay. The larger vessel quickly found itself unable to recross the bar into deep water and ran aground just as the sea withdrew until the next tide—plenty of time for a mob to be roused on shore, which then waded out and set the ship on fire and took its British crew hostage.

Boston itself has expanded seaward since the eighteenth century, and only two of the original colonial wharves survive. Griffin's Wharf, where the *Dartmouth*, the ship that became the focus of Bostonians' anger during the Tea Party, was moored, is long gone, and its precise location is a matter of some uncertainty even on maps of the period. I give up my search and make instead for the Boston Tea Party Ships & Museum. The staff here aren't about to let any niggling doubts about historical authenticity hold them back. Their ships are replicas, though they are at least afloat in the harbor.

The costumed crew haven't got the idiom quite right either.

"Boston Harbor a teapot tonight!" shouts one.

"I'm really sensing the anger there, Joseph," another returns.

There is no photography permitted in the museum: "We don't

want it leaking out on ye olde Book of Faces," I'm told. "King George checks that all the time." One of only two genuine surviving tea chests is preserved on a rotating plinth behind a cylinder of glass like a religious relic.

But what was the "Tea Party," and how did the tide play a role? On the night of December 16, 1773, a band of fifty or so "Sons of Liberty," some disguised as Mohawks, rushed aboard the moored ships of the East India Company and dumped overboard more than three hundred chests of tea. On my historic tour, a Hispanic boy aged about eight is having great fun reenacting this part of the action over and over again while his parents listen seriously to the commentary on deck. This disobedience was met with fierce British repression and prepared the way for the Revolutionary War two years later.

In seizing their moment—it was drawing toward the end of the last day on which the *Dartmouth* was permitted to remain in harbor before either the duty was paid or she was forced to sail without unloading—the Bostonians missed a trick and nearly lost the initiative. For, according to Bruce Parker, a former chief scientist of the US National Ocean Service, "one aspect of that famous event did not go quite as planned." At the moment when the tea was tipped overboard, the tide was not only low, but exceptionally low, it being a spring tide, and a spring tide moreover when the moon was at perigee, at its closest to the earth. The *Dartmouth* was aground in the mud with perhaps only a couple of feet of water lying around her. Her cargo of tea had nowhere to go. As chest followed chest, they simply piled up next to the ship. In desperation, the rebel commanders directed some of their number to wade in and do what they could to disperse the tea, breaking up the chests, and batting the loose tea into the water with oars. Undamaged tea chests might be retrieved; only broken chests mixed with seawater would render the disputed tea valueless.

When the flood tide came later that night, the tea was still there, and it was not until the ebb in the wee hours that it at last began to be transported away on the water. The perfect coup de theatre for the protesters would have been for the British to find the cargo of tea simply vanished from the ship in the morning. If the dumping had happened at the top of the high spring tide, the tea would have been spread far and wide through Boston Harbor within hours, and most of it never seen again. As it was, the dumped tea had insufficient time to disperse, and large rafts of it washed up on nearby shores south to Boston Neck and across the harbor channel on the Dorchester flats. Many Boston citizens awoke with joy to find the tea strewn along the shore, providentially deposited there in the night by the tide. But a few took surreptitious advantage of the flotsam. Ebenezer Withington found half a chest in the marshes, took it home, and began to sell it until some Sons of Liberty seized it from him and burned it on Boston Common. Others took to the freezing water in canoes to retrieve a little of the precious commodity for themselves.

Later, I ask Benjamin Carp, a historian of the period at nearby Tufts University and a contributor to the Boston Tea Party Ships & Museum, what he makes of these events. "Many of the tea destroyers worked on the docks or at sea," he points out. "They probably knew the timing of the tides. But they had to do the best they could with the time they had, before the imperial government or someone else interfered. I wouldn't say that the tides foiled the Bostonians' plans. The low tide just added an element of humor."

SUR LE FLUX ET REFLUX DE LA MER

It may have escaped the American revolutionaries' notice, but the tides were by this time understood in new mathematical detail. With English science such a nest of vipers at the beginning of the eighteenth century, and the mutual enmity among its leading lights—Newton, Hooke, Halley, and Flamsteed—needlessly holding back progress, it is perhaps not a surprise to learn that the next major developments in the theory of tides would come from elsewhere. The attention of the navy was, in any case, drawn toward a possibly more tractable problem in navigation.

In October 1707, four ships in the fleet of Admiral Sir Cloudesley Shovell, returning from the Siege of Toulon during the War of the Spanish Succession, ran aground on the Isles of Scilly, with the loss of more than fourteen hundred men. Attempting to allow for stormy weather pushing them off course, the ships' navigators had reckoned they were still in deep water far to the west of land. If they had been able to calculate their longitude, they would have known they were, in fact, farther east and would have altered course to avoid the rocks. The disaster led the British government to establish the Board of Longitude to encourage the development of a device that would enable ships to know the precise time and thereby to work out their position with greater accuracy. Of course, had Sir Cloudesley's ships run aground owing to an error in calculating the tides, it is interesting to speculate whether greater effort would have continued to be directed toward that problem.

Meanwhile, the French had begun a program of systematic measurement of tides at their major ports. Understanding of the physical principles involved was hampered, however, by loyalty to the outdated theory of Descartes. In 1720, the French Royal Academy of Sciences had taken the visionary step of instituting an annual essay competition. In an effort to raise standards and broaden the perspective of French science, the competition was international and open only to nonmembers of the academy. The subject of the essay for 1740—announced in good time in 1738—was *Sur le flux et reflux de la mer*.

Four winners were chosen, reflecting a brief flowering of scientific internationalism before much of Europe was overtaken by revolution and nationalist jealousies. They were: Colin MacLaurin, the professor of geometry at the University of Edinburgh and a protégé of Newton who had investigated rotating fluid bodies and the gravitational forces governing ellipsoids, the geometric shapes that most accurately approximate the shapes of bodies in the solar system; the Dutch-born polymath Daniel Bernoulli, who held chairs in mathematics, physics, anatomy, and botany at Basel; Leonhard Euler, Swiss-born but at the time a professor of mathematics at St. Petersburg, to which position he had been recommended by Bernoulli, who had himself spent a few years in the Russian city. (Both Bernoulli and Euler were greatly interested in the behavior of fluids, and both had been already much garlanded by the French Academy.) The fourth winner has been treated less kindly by the passage of time; he was Antoine Cavalleri, a French Jesuit scholar at the University of Cahors, whose essay patriotically labored to explain the tides using Descartes's old concept of vortices working through a universal ether, and rejected the new English theory of gravity.

More au fait with Newton's work, each of the other three men was able to make small but significant advances in tidal understanding. Euler confirmed that it is the horizontal force alone acting on each little mass of seawater, and not the vertical pull of gravity, that gives rise to the tides. This idea would be taken further by the great French mathematician, Pierre-Simon Laplace, at the end of the century. MacLaurin proved what Newton had merely assumed: that the tidal bulges on the sides of the earth facing and opposite the source of gravitational attraction give the simplified model of the ocean-covered earth the shape of a prolate spheroid.

Bernoulli extended this geometric analysis to calculate the departure from a perfect sphere of the ocean-covered earth with the sun and moon in any relative positions, taking into account distortions caused by the orbit of the earth about the sun and its rotation about its own axis, and even the mutual gravitational attraction between the solid earth and the oceans. Bernoulli furthermore showed that Newton's ideal theory could be reconciled with the tide experienced in real places, provided that long series of tidal observations were made. This was "the first practically reliable advance in prediction technique since the crude rules of thumb of the 13th century," according to David Cartwright, a leading authority on tidal oceanography.

Nevertheless, this is hardly the ultimate destination that any self-respecting theorist would hope to reach. The theorist would like to be in a position to calculate the tide for any given place on earth at any given time *ab initio*, without reference to empirical local data, but with the help solely of Newton's laws and calculation involving all the bodies that exert a gravitational force on the earth and its oceans. Yet the necessary working approximation adopted by these mathematicians of an earth evenly covered with water was obviously a pretty poor model of the reality. You only had to stand on the shore to see that.

The poet Alexander Pope must have sensed the impasse that had been reached. His *Essay on Man*, published in 1734, includes the famous couplet "Know then thyself, presume not God to scan; / The proper study of mankind is man." The following verse adds these gently sarcastic lines:

> Go, wondrous creature! mount where science guides,
> Go, measure earth, weigh air, and state the tides;
> Instruct the planets in what orbs to run,
> Correct old time, and regulate the sun;
>
> . . .
>
> Go, teach Eternal Wisdom how to rule—
> Then drop into thyself, and be a fool!

THE FRENCH NEWTON

The mathematician, astronomer, and physicist Pierre-Simon Laplace—who would come to refer to himself as the "French Newton"—was born a few miles from the Normandy coast in 1749. Wisely avoiding taking sides during the French Revolution and after, he was elevated to the peerage by Napoleon, who shared his passion for mathematics and eventually appointed him interior minister, and then elevated still further, to marquis, by Louis XVIII when the monarchy was restored. Laplace's greatest achievements were in developing the theory of probability, which led him to consider whether the solar system could have evolved by chance, and thereby to a deeper interest in astronomy. His multivolume masterwork was the *Traité de mécanique céleste* ("Treatise on Celestial Mechanics"). But early in his career, he was drawn to the problem of the tides.

Using sophisticated mathematical methods of his own devising, Laplace was able to supersede the theory of Newton with a new "dynamic" theory that, for the first time, made allowance for at least some of the conditions imposed on tidal movement by the oceans as they actually exist. Newton had treated the oceans as a system in equilibrium, which is then disturbed by the gravitational attraction of the moon and sun. That is to say, if the moon and sun were not there, Newton's seas would just sit there, statically wrapped around the earth. Laplace's dynamic theory made no such dangerous assumption. For him, the oceans were *already* in motion. They could be regarded as oscillating basins, their contents caused to slop about at various natural rates by the rotation and orbit of the earth and damped by friction between the water and the earth. This development successfully united Newton's theory of gravity with Galileo's earlier intuition about tides, based on the movement of water in his Venice barge.

Laplace's equations made use of the law of conservation of mass (no water is created or destroyed) and the law of conservation of momentum (moving water keeps moving in the same direction unless acted upon by a force) applied in two horizontal directions (which might be latitude and longitude). They showed how the forces of gravity, friction, and retardation owing to the water's inertia act on each parcel of water—my cubic centimeter, say—and yielded figures for its displacement both vertically and in the plane of the sea surface. Such a parcel of water in the North Sea, for example, might make a typical "tidal excursion" of some 5 kilometers but rise and fall by only a meter or two.

The Frenchman's new ideas were outlined in a pair of papers published by the French Royal Academy of Sciences in 1775 and 1776. He managed to explain why the two tides each day are approximately the same height, whereas Newton's calculations had suggested they should be relatively unequal. Laplace's analysis also confirmed that the vertical force on any given mass of water is very small, and that it is the far greater horizontal force that makes the water move and gives rise to the tides. There is no horizontal force at all near the equator, where the moon is directly overhead; the water here is already as close to the moon as it can get. This force instead reaches its maximum at an angle of 45 degrees off the line of gravitational attraction between the earth and the moon, which explains why the regions of the earth that experience the greatest tides tend to be in temperate latitudes, north of the Tropic of Cancer and south of the Tropic of Capricorn.

Laplace's equations told him other surprising things. For example, in deep waters the high tide is in phase with the gravitational force causing it, but in shallow waters the reverse is true, meaning that the actual tide is out of phase with the presence in the sky of the moon that causes it. In other words, the tidal bulge raised by the moon "follows" the moon as the earth rotates beneath it—until it runs into a coast, when complications arise. Laplace showed that the progress of this bulge or global tidal wave (not to be confused with the phrase "tidal wave" inaccurately applied to tsunamis) was related to ocean depth. At a time when many people still assumed the seas were deep beyond measure, and others simply reasoned they must be as deep as the mountains are high, Laplace bravely put a figure on the oceans' average depth of 4 leagues, or 12 miles. This sounds large—it is indeed an overestimate of what we know today—but it is still a very small distance compared with the radius of the earth. The ocean is like the thickness of a pencil line scribed by a compass, or, as I have seen it imaginatively described, like the film of dew that coats your car on a cold morning. Seen like this, it is easy to understand that the movement of the ocean's waters must be predominantly horizontal rather than up and down.

For all the mathematical virtuosity he was able to deploy, Laplace was still forced to make simplifications. That the oceans were uniformly shallow (at 12 miles deep!), a premise that allowed him to neglect certain complicating motions expected in very deep oceans, was not his only assumption. He also assumed that the bottom of the ocean was rigid, which it isn't, since it is subject to its own deformations owing to gravity. Like Newton before him, Laplace returned again and again to the tides throughout his long and productive career, characterizing it as the "thorniest problem in physical astronomy." The statement was, in effect, a confession that the tides were truly far more complicated than scientists had feared.

The opportunity to put Laplace's equations to work in making real tide predictions would not come until the invention of powerful calculating machines able to manipulate the large volumes of data involved in the late nineteenth century. Nevertheless, they are still the theoretical basis of all modern tidal predictions—although, in a reversal of the suspicion the French had once shown toward Newton, it took the British until well into the twentieth century to start using Laplace's superior methods.

VOYAGES OF DISCOVERY

Laplace's tidal equations were all very well for a planet evenly flooded with water. But we do not live on such a planet, and the elegance of the Frenchman's mathematics were no match for the complications of actual oceans and coastlines. It remained imperative, therefore, to continue gathering data. The great voyages of scientific discovery of the eighteenth and nineteenth centuries made it part of their mission to measure the tides wherever they went. In some places, these observations seemed to contradict Newtonian expectations. Explorers found locations where there was only one tide a day, or no discernible tidal movement at all. Captain James Cook made measurements in the South Pacific at the request of the Astronomer Royal, but he could not square them with existing theory. Often, it seemed, unscientific seafarers' lore served better. Late one June evening in 1770, Cook's ship, the *Endeavour*, ran aground on a coral reef off the coast of New Holland, near what is now Cooktown in Queensland, Australia.

> At this time I judged it was about high water, and that the tides were taking off, or decreasing, as it was three days past the full Moon; two circumstances by no means in our favour. [The ship could well be stuck for up to a fortnight until the next spring tide.] As our efforts to heave her off, before the tide fell, proved ineffectual, we began to lighten her, by throwing over-board our guns, ballast, &c. in hopes of floating her the next high-water; but, to our great surprize, the tide did not rise high enough to accomplish this by near two feet. We had now no hopes but from the tide at midnight; and these only founded on a notion, very general indeed among seamen, but not confirmed by any thing which had yet fallen under my observation, that the night-tide rises higher than the day-tide.

The French explorer Louis-Antoine de Bougainville reported similar oddities in his *Voyage autour du monde* ("Journey around the World") of 1771. His account is celebrated for its alluring description of the Polynesian island of Tahiti, which he called New Cythera. Here, the expedition's naturalist, Philibert Commerçon, who named the glorious flowering bougainvillea in honor of his captain, was revealed to have brought his mistress along on the voyage disguised as his valet; thus did Jeanne Baré become the first woman to sail around the world! The island is a sensuous paradise indeed, but the tide made it a dangerous trap. Leaving her anchorage, Bougainville's ship, the *Boudeuse*, had just successfully steered outside the reef when the wind died, and "the tide and a great swell from the eastward, began to drive us towards the reefs." Had the mishap occurred inside the reef, Bougainville could not help but think, "the worst consequences of the shipwreck . . . would have been to pass the remainder of our days on an isle adorned with all the gifts of nature, and to exchange the sweets of the mother country, for a peaceable life, exempted from cares." But adrift on the ocean side of the reef, with waves crashing on the sharp coral, the crew would certainly be drowned or dashed to pieces. Fortunately, the ship's longboats were able to tow her out of danger.

Bougainville also made several stops in the Falkland Islands, or *Îles Malouines*, which he planned to claim for France and colonize with French-speaking settlers driven out of Canada by the British. On one occasion, he observed the tide there rise and fall three times within a quarter of an hour "as if shaken up and down." This was probably a transient effect due to an undersea earthquake; a similar phenomenon was reported in December 2004, when local people noticed the water rise rapidly and saw ducks "turbo jetted" from its surface, and linked it to the devastating Sumatra-Andaman tsunami which happened more than 10,000 miles away.

But against the growing volume of tidal measurements now coming in from explorers and ports around the world, such anomalies were increasingly easy to spot. Of far greater significance was the mass of routine data that would make sea trade safer and more profitable than it had ever been before.

THE PRICE OF TEA II

The link between tide and trade is spectacularly illustrated by the annual race to bring the freshest new season's China tea to eager vendors in Victorian London. Tea became highly fashionable in the

eighteenth century. From 1853, when Fouchow (Fuzhou)—closer to
the tea groves than Canton (Guangzhou)—joined the list of Chinese treaty ports open to international trade, fast sailing ships known
as tea clippers set out to see which would be first home with its luxury
cargo. The owner of the winning ship stood to make a substantial
profit because his tea would command a premium price. Londoners
and sometimes even the ships' crews themselves bet large sums on the
outcome. "For the general public, this was like following the Derby,"
according to Jane Pettigrew, writing in *A Social History of Tea.*

The informal racecourse took the clippers down through the South
China Sea, across the Indian Ocean, around the Cape of Good Hope,
and then north catching the trade winds up the Atlantic Ocean,
before turning up the English Channel and through the Strait of
Dover to the entrance of the Thames estuary. Winner and loser were
often decided on this last leg, since a ship arriving with a six-hour lead
might find all was lost if it had to wait out a foul tide before continuing into port.

The most exciting of these contests took place in 1866. The *Taeping,*
the *Ariel,* and two other clippers were lying neck and neck when they
were spotted at the Azores three months after leaving Fouchow. As
they raced up the Channel, the *Taeping* and the *Ariel* pulled clear of
the others. By dawn on September 6, after a journey of ninety-nine
days and 14,000 miles, the *Ariel* was leading by a matter of minutes,
with the *Taeping* gaining on her, as the two clippers approached the
Thames estuary, where they were obliged to pick up tugboats to lead
them safely into dock. Here, it was the *Taeping* that had the luck to
pick up the faster tug; she arrived fifty-five minutes ahead of her rival
at Gravesend, where both vessels now had to contend with the tide.
The *Ariel* had to wait until evening for the tide in order to enter the
East India Dock, while the *Taeping,* with a shallower draft, was able
to continue upriver to its dock. The *Taeping* claimed the premium of
10 shillings a ton for delivering her cargo first, but honorably shared
it with the *Ariel.* The two ships' captains also shared their personal

£100 prize. But the arrival of the two large clippers hot on each other's heels—and shortly followed by a third—had an unfortunate consequence: it meant there was almost immediately a glut of the new season's tea on the market, and so the premium system was dropped in future years. On this occasion at least, it seems, market forces proved stronger than tidal forces.

PICTURING THE TIDE

While scientists struggled to put their new understanding of the tides to practical use, painters struggled to represent the existence of the tide at all. It is hard to capture in a single image the sense that the earth's oceans are constantly rising and falling. The mesmeric nature of tidal movement understandably confounds the artist limited to a single canvas. The genre of art known as marine painting is full of churned-up seas, and ships of all kinds at sea, fighting the waves or each other. Vessels under sail are highly picturesque, and the heroic struggle between humans and the elements could hardly be more clearly represented than by placing such a craft amid a chaos of waves. But in these scenes there is no fixed point by which the state of the tide may be known. Quayside scenes of ships loading or unloading provide this comparison, showing, as they do, the relation of the water to the land. Undoubtedly, these views were also a more attractive proposition to many artists, who could paint their subject while avoiding the discomfort of an actual sea voyage. In these paintings the tide may be high or low—more often low, in my survey, as it seems artists find greater visual interest in an exposed foreshore, whereas the rest of us find high water prettier, or at least less smelly. But there is still nothing in these compositions from which the viewer can deduce that these waters rise and fall.

My favorite among the few works I have found that do manage to convey a sense of the tidal cycle is a large canvas by George Vincent called *Dutch Fair on Yarmouth Beach*, painted in 1821. It shows several handsome Dutch sailing ships resting far up on the dry beach. Around them, stalls have been set up, with townsfolk in their finery, all on the sand—on what, in a few hours' time, will in fact be the seabed. This is the trick of the painting. These ships are too large to have been manhandled onto the sand so far from the water. It is clear that the tide has deposited them there, and it is the tide that makes the trade possible, its flow first bringing in the shipboard goods, its ebb then permitting the townsfolk to crowd around to buy. The vessels' sails remain hoisted even while they are on the sand, anticipating that the market will soon be over and the tide will return to float them off.

Vincent portrays a real event in a real place, an occasion also favored by his fellows in the Norwich School of painters, John Sell Cotman and John Crome. The Dutch fair took place on the Sunday before the autumn equinox when the short herring-fishing season began. We can tell that it is, indeed, at Yarmouth because the artist has been careful to include the town's monument to Lord Nelson—an earlier and finer column than the one in Trafalgar Square. Yarmouth's historical prosperity as a fishing port owes much to its nearly flat, sandy beach, which permits boats to be beached in this way when the weather is kind. It was certainly not always a safe shore. Other paintings by these artists are more inclined to melodrama. They show rescues from ships that have been wrecked on the dangerous sandbanks that run parallel to this same beach. Many were commissioned by George Manby, the master of the Great Yarmouth artillery barracks, in order to demonstrate his own invention, the Manby mortar, which could fire a line to a ship in trouble off the coast and so bring stricken sailors to safety.

Writers were equally nonplussed. That the tide anywhere on earth became both calculable and comprehensible in fundamental physical terms was no help in storytelling. For them, the chief narrative merits of the tide are its life-threatening potential and its inexorable cycle—

and not, it seems, its largely knowable times and its knowable reach. The tide is involved with greater literary artifice in other works, as we shall see. But these examples seldom require us to have any direct experience of the phenomenon. Only rarely is an inkling of scientific understanding indicated.

William Golding's terrifying novel *Pincher Martin* describes a drowning man's last days—or maybe only moments—on a tiny ocean rock. The stranded naval officer Martin quickly realizes that his island prison is "land only twice a day by courtesy of the moon. It felt like solidity but it was a sea-trap." Delirious with starvation, he sees the tide's tricks, such as the illusion that his rock is cleaving purposefully through the ocean, whereas it is, of course, the ocean that surges past the rock. Uncertainly, he remembers, "The tide is a great wave that sweeps around the world—or rather the world turns inside the tide, so I and the rock are—"

Hastily he looks down at the rock between his feet.

"So the rock is still."

Still or not, the rock—if rock it ever was—is his grave.

More conventional is Victor Hugo's use of the tide to bring death to his character Gilliatt, a fisherman on the Channel Island of Guernsey, in *Toilers of the Sea*. When the island's first steamship is wrecked on rocks, the owner is desperate to recover the engine and promises his niece in marriage to whoever can salvage it. Gilliatt takes up the challenge, laboring for weeks on the rocks, fighting off storms, loneliness, and even an octopus, in order to free the engine from the wreck. He is ultimately successful but returns to port to find the girl promised to another, and honorably refuses his prize, saying he does not love her. In despair, he takes himself back out to the rocks where he can watch the newly married couple sail away, seated on a natural rock throne that is covered at high tide.

The melodrama of the romantic tale is in sharp contrast with Hugo's almost clinical description of the geology and the sea in this alien place where the tide rises calmly, "not by waves, but by an imperceptible swell." In strikingly unimmersive prose, Hugo simply notes the passing of time by the height the water has reached on Gilliatt's body. The novel ends, "At the moment when the vessel vanished on the line of the horizon, the head of Gilliatt disappeared. Nothing was visible now but the sea."

Some authors are determined to show they know more about the tide than their characters do. For example, Daniel Defoe uses the tide to reduce Robinson Crusoe in an instant from a civilized man to the appearance of savagery. Early on in the story, the flood tide brings Crusoe's wrecked ship close to shore, where it runs aground, so that when it recedes, he is able to wade out to the vessel and retrieve some provisions. "While I was doing this, I found the tide began to flow [again], tho' very calm, and I had the mortification to see my coat, shirt, and wast-coat, which I had left on shore upon the sand, swim away," Crusoe relates. It is a naïve mistake for such an experienced sailor to make. Surely, Defoe is having a joke at his character's expense, and helping us to identify with Crusoe, because this is precisely the sort of error that we landlubber readers might make. Once he has grown accustomed to island life, however, Crusoe lives in natural harmony with the tides and is able to use their ebb and flow to assist the circumnavigation of his domain in the canoe he has made.

In *Kidnapped*, Robert Louis Stevenson also has fun with the tide at the expense of his central character. Like Crusoe, David Balfour is marooned on the tiny Scottish island of Earraid after his vessel has run aground. In the days that follow, he makes several desperate circuits of the island but can see no escape. At one point, some fishermen sail heartlessly past, laughing at his antics. In fact, he has simply been unlucky in unknowingly timing his walks to coincide with the high tide. Finally, on the fourth day "of this horrible life of mine," a boat hails him to apprise him of his error, and he learns that the island is linked to the neighboring larger island of Mull by a causeway visible only at low tide. "A sea-bred boy would not have stayed a day on Earraid," he is forced to admit to himself as he makes his way sheepishly back to safety.

Even full-blooded maritime tales do not make as much use of the tides as you might expect. Because the tides are of little consequence on the open main, they do not appear much in Frederick Marryat and Jack London, Melville and Conrad, for example. In nautical narratives confined to coastal waters, however, it is occasionally a different story, and for perhaps the most gripping and sustained impression of the tides in fiction, we must return to the north German coast and the extended estuary of the River Elbe.

Erskine Childers's 1903 adventure, *The Riddle of the Sands*, is less read now than it was, and it contains some unappetizing British paranoia about the Continent. But in its watery immediacy, it is hard to surpass. Carruthers, suffering a bout of ennui at his desk in the Foreign Office, seizes on the mysterious invitation of his acquaintance, Davies, to join him on a sailing expedition to the Frisian Islands. Here, the pair uncover a secret site where the Germans are preparing for an invasion of Britain.

The details of the plot are not important. The delight of Childers's story is that it places you so close to the water that you can almost feel it trickling through your fingers as you trail your hand over the side of the boat. As the story gathers pace, Childers inserts a footnote: "I exclude all the technicalities that I can, but the reader should take note that the tide-table is very important henceforward." As if you would never be without one! The reader soon finds that the skills required to navigate these shallow waters, where the topography is constantly changing and the harbors and quays are functional for only a couple of hours in each tidal cycle, are very different from those needed for an ocean voyage. Charts are of little use, and even safe-channel markers may lead the unwary into danger in the ever-shifting intertidal landscape. With no conspicuous landmarks all around the flat horizon, it is easy to lose your bearings completely. Better to use the tides to feel your way because the tides run strongly in the channels, but hardly at all over the shallows. Sometimes, feeling your way becomes literal: a boat that does not draw much water can, if the helmsman is careful, "tide over" a sandbank by inching forward as the sand grinds gently against the keel. Our expression "touch and go" may also originate from this kind of passage making, although it has long since lost its nautical flavor.

The climax of the Englishmen's escapade is a long rowing expedition under cover of darkness—and the improbable pretense of being on a nighttime duck shoot—from Norderney Island west to Memmert Island through channels sheltered by the barrier island of Juist from the North Sea. The journey is only accomplished at all because Davies and Carruthers have been able to use the tide to their advantage, crossing the shallowest point, the "watershed," at the top of high water, and using the flow and ebb on either side of this obstacle to speed them along. "What a race it was! Homeric, in effect; a struggle of men with gods, for what were gods but forces of nature personified?"

8

A PLACE
OF RESONANCE

A BIT OF A BORE

We have come to see the bore. There are about forty of us in all, from across Canada, from the United States, a few from Germany and Britain and perhaps elsewhere. We are standing on a high pier built partway out over the Shubenacadie River on the supports of an old bridge. These stone bases, which stretch on across the river, bear the scars of a century's abrasion by its fast-moving, sand-laden waters. A replacement bridge just upstream has a sharp dip in the middle because the foundations have settled badly in the constantly moving sand beds. It is said the engineers ignored local advice when they designed it forty-odd years ago.

The Shubenacadie River empties into the Minas Basin, which forms the southern branch of the Bay of Fundy in Nova Scotia. The bay has the highest tides in the world. The bore wave surges up the river when the tide comes in. Some of us may simply have been lucky with the date, but most, I suspect, have chosen as I have to make this rendezvous with nature on the day of the highest tides of the year.

A full moon last night has increased my sense of anticipation for the phenomenon that lies in wait for me. Driving out to the spot, I passed tidal inlets very different from those of East Anglia with their smooth mud. These channels are vigorously carved out of sand and

sandstone, steep-sided and sharply corrugated like miniature ravines, with clear signs of the regular torrential movements of water. There is something desperate about the way the grasses on the bank tops are tethered to the soil. On the outer edge of some of the bends in these inlets, they lie still slicked down, sodden and matted with mud, from when the last tide washed over the edge hours ago.

We are early—we all know the tide does not wait for man—and an expectant babble has started up. People philosophize about the fixity of time and wonder how quickly the spectacle will pass. Locals compare the merits of different viewing spots. A woman and her daughter are discussing strategy—who will take photos and who will shoot the video. I soon learn there are tidal bores, and there are also tidal-bore bores. "This is not very impressive, this one," says a man. "Truro is best. But this is fine." Already, though, chocolate-colored water is swirling at the bridge supports, making whirlpools the size of vases. "This is the highest one in a while," another man rejoins. "The perigee is right. The earth's tilt," he adds, wrongly.

Downstream, the river widens in a broad, shallow band with sand-banks stretching almost the whole way across. A bald eagle alights in a Scotch pine. The bore is said to be ten minutes "behind schedule." (Although the times of high and low tide can be predicted in principle with great accuracy, local weather effects can hasten or delay their arrival.) The eagle flaps lazily to the other side of the river. The light breeze blowing up the valley seems to stiffen a little. But the river is still emptying. New sandbars are emerging as it does so, their arrival advertised by a last flurry of surf as the water falls away. Three eagles are now on station inside the bend. They come to catch striped bass flung up by the bore wave. How do they know it is coming?

THIS IS NOT MY FIRST BORE. The best-known bore in Britain is on the River Severn. The river has the right funnel shape to focus the energy of the tidal wave that travels in along the Bristol Channel. I make the journey across the country to see it in the wee hours one Sunday morning in February when the Environment Agency has issued storm surge warnings for the whole Severn valley south of Gloucester.

It is just dawn when I arrive, and it is apparent that the event is quite an attraction. Cars are pulled up untidily along the shoulders, and people line the bridges over the river in expectation. At the Minsterworth church, one of the most popular vantage points, I join about five hundred spectators. There must be thousands here in all. Photographers fiddle with their tripods. A helicopter hovers overhead. Many people are standing right at the water's edge. A man with a local burr berates them—"I'm going to give you a clue. That's the water level, up there"—and points over their heads higher up the bank to where a line of broken reed stems and other plant matter was deposited when the previous bore ran through. He stalks off, shaking his head.

What I see that morning is not the foaming wave I had expected, but something much odder. Gradually, the river appears to buckle, catching the low sunlight like a shining sheet of metal as it concertinas into sections, each with a clear gradient up or down. The transformation has the sense of a great mass behind it like a glacier subjected to some huge unseen stress. At first, there are several of these giant

202 • THE TIDE

undulations, and the surface of the water remains unbroken, a series of rolling hills suddenly turned to liquid, an ocean swell injected into the tranquil setting of a willow-banked stream.

There is something deeply uncanny about this water, still bounded, yet plainly out of place: water that is not level, that has lost its balance. Patricia Highsmith senses the extreme weirdness of sloping water in her short story "One for the Islands," in which an ocean liner finds itself sailing steadily downhill—"at about a twenty degree angle with the horizon, and such a thing had never been heard of before, even in the Bible"—and the passengers understand they are, in fact, making their last voyage in life. But on the ocean, there would be no reference point for this; you would hardly know you were sailing downhill until the angle became quite steep. On land, there is a reference point—the horizon. Water must make a horizon—that is what it does under the influence of gravity—so has the entire land tilted? Our whole horizon is rocked as the wave passes.

As this disturbance advances up the river, the waves appear to bunch up and gain height. Then, all of a sudden, the first of them crests and is upon us, breaking through the reeds as it sweeps past, causing some of the spectators to step smartly up the bank. The wave is odd too. It passes, but the water does not go back to its original level. In fact, it doesn't go down at all. It stays at the new height, up by almost a meter in a moment, and then continues to rise rapidly through the grass and mud. Behind the wave, which now pushes on toward Gloucester, the river water is left swilling around as a single mass of fluid cupped in its bed like soup in a bowl, slopping at the edges, lurching from bank to bank, inducing further vertigo.

THE SHUBENACADIE BORE will surely be very different. There is an un-English grandeur about the scenery. Apart from the bridges, the countryside is completely wild. This river is vast and shallow, a thing of raw force, but we are high above it, out of danger, and unlikely to get our feet wet.

At last, somebody spies ripples advancing across the sandbanks far downstream. They are very small, but turbulent, fighting their way across the top of the water that is still trying to leave the river. Then the wind increases and the temperature drops sharply. Something is coming. The wave begins to build in volume, raising a chop where it meets the old river water, and forming rapids across the sandbars. On the far side of the river, a vigorous eddy forms as the flood tide tries to barge its way past the last of the ebb. The guide at the interpretive center is keen to manage our expectations. The shifting sands alter the bore wave. "Last year," she says, "the highest bore we had was this big." She holds up a finger and thumb an inch apart. We laugh.

Today, the bore will be more respectable, though no record breaker. As the advancing wave begins to sweep over the sandbanks, I notice there is now a visible gradient on the surface of the water. The difference in the height of the water between the upstream and downstream sides of the bridge supports is a foot or more. Below us, the water is flowing fast and straight up the middle of the channel. The breeze has picked up too. Standing on our platform now feels like being in a wind tunnel.

Then come the rafters, waterproofed thrill seekers in powerful inflatables. But they don't look thrilled beneath their synthetic layers. The boats buzz around, disconsolately looking for the biggest part of the wave to ride. Their engines shatter the atmosphere, and we begin to disperse with a sense of mild anticlimax. It has been natural, but not all that exciting—an authentic Canadian experience, I think, unkindly. The eagles have watched it all and not made a move.

Afterward, it occurs to me that what I really witnessed is not the wave on its own, but a concerted event in nature. The temperature, the wind, and the eagles, as well as the river and the tide, all conspired to orchestrate a "happening," a significant occasion that requires a special place and a special time like a ritual. Suddenly, the old northern mythologies make more sense, with their gods and goddesses of ocean and river, with their symbolic birds and sea creatures, all interlocked in stories of struggle and conquest, bargaining and cooperation. These deities have enacted the ancient narrative for me. They live among us still.

The sea is too large a portfolio for one god. Neptune was designed for the Mediterranean Sea. He would be out of his depth in the Atlantic Ocean beyond the Pillars of Hercules. Better the Norse pantheon, which assigns its sea gods and goddesses separate duties. Jörmungandr is the sea serpent so large that he encircles the earth, creating the ocean currents. Rán is the destroying sea goddess who collects sailors' voluntary and involuntary offerings in her net. She accepts both their propitiations made before a voyage and their treasure if their ship should founder; jetsam and lagan are her spoils. Rán is married to the giant Aegir, giver of banquets, who represents the abundance of the seas. They have nine daughters, the billow maidens, each one of whom represents a different kind of wave—the roller and the breaker, the transparent wave, the foamy wave, and the wave that makes the boat pitch. And any of these is capable of being stirred to heights of fury by the shape-shifting Loki. Loki sometimes takes the form of a salmon, while a nameless eagle perches on the world ash tree, Yggdrasil.

In Shinto mythology, the tide is controlled by two fiery jewels kept by the dragon king of the undersea world, Ryūjin. Manju controls the flood; Kanju, the ebb. It is said that the jewels were the size of apples and shone with a fiery light like dragon's eyes, although I find it hard to shed the modern image of colored traffic lights.

It is also said that the empress Jingū, widow of the fourteenth mikado of Japan, Chūai, used the tide jewels in her conquest of Korea around 200 CE. Her husband refused to believe in the land across the sea that the gods had promised to her in a dream, and so after they destroyed him, she built a fleet of ships and set out westward across the Sea of Japan to see for herself. As they neared the unfamiliar land, Korean warships sailed out to intercept them. Jingū took the jewel of the ebb tide, Kanju, from her girdle and cast it into the sea. The sea instantly withdrew, leaving the Korean vessels stranded. Thinking their enemy would be similarly stuck, the Koreans promptly disembarked and rushed across the sands to attack. As they approached, however, Jingū threw in the flood jewel, Manju, and the sea returned, drowning the armored Korean defenders, and enabling the Japanese to sail up to the coast and claim the land.

The Mi'kmaq people native to Canada's Maritime Provinces have a myth specific to the great tides in their region. One day, the provider deity Kluskap summoned Beaver to create for him a bath, and Beaver made a dam across the Minas Basin, stopping the flow. Whale demanded to know why his water had been taken, and so, not wishing to enrage Whale, Kluskap stepped out of the bath, and Whale broke down Beaver's dam, releasing the trapped water out to sea.

Recent underwater surveys in the Bay of Fundy have intriguingly suggested that this story may not be a way of thinking about every tide, but actually an account of a single cataclysmic event. The Minas Basin is like a barbed arrow at the head of the bay. At its narrowest point, between Cape Split and Parrsboro, the channel linking these two large bodies of water is only 5 kilometers wide. Here, scientists believe they may have found the remains of a stone causeway or "dam-like structure" built during the earliest human occupation of the area, when sea levels were lower. Changes in the climate since then may have engulfed the causeway, allowing the sea to rush into the basin. Evidence from tree remains and oyster shells indicates that the Fundy tides may have had a range as low as 2 meters ten thousand years ago—too modest to raise a bore on the Shubenacadie. With rising sea levels and an increase in volume of seawater in the whole bay, the tides would have greatly increased. The mythmakers seem to have kept pace with these natural changes, for they also have a story specific to the tidal bore, in which the lobster of the sea is called to battle against the eel of the river, who is pushing out mud into the clear bay. The battle is so violent that, to this day, great waves are sent up the river and the sea is permanently discolored.

IN THE LARGE RIVERS on the east coast of England once invaded by Vikings, the bores are still named aegir, or eagre, after the Norse sea giant. The word "bore" itself comes from the Old Norse *bára*, meaning "wave." The most striking of these occurs on the River Trent in Lincolnshire. George Eliot reveals the deep connection between this land and the mythology in *The Mill on the Floss*. Living by the Floss, young Tom and Maggie Tulliver wander its banks "with a sense of travel," going to see "the rushing spring-tide, the awful Eagre, come up like a hungry monster, or to see the Great Ash which had once wailed and groaned like a man." Eliot's contemporary, Jean Ingelow, born in Boston, notes the devastating effect of the bore as it floods the town in her poem "The High Tide on the Coast of Lincolnshire (1571)." As the church bells ring out in alarm, the seawall is breached and the fields are flooded. Then,

> A mighty eygre reared his crest,
> And uppe the Lindis raging sped.
> . . .
> So farre, so fast the eygre drave,
> The heart had hardly time to beat,
> Before a shallow seething wave
> Sobbed in the grasses at oure feet:
> The feet had hardly time to flee
> Before it brake against the knee,
> And all the world was in the sea.

Ingelow's ballad is pure Victorian melodrama, complete with boats hoisted into the marketplace and people taking refuge on rooftops, but it is also the product of accurate observation, as she conflates the historic flood of 1571 (there were indeed devastating tidal surges along the east coast in 1570 and 1571) with a more recent one, in 1810, when a milkmaid lost her life. Ingelow's preamble includes the line "Men say it was a stolen tyde"; in other words the flood came on top of the previous flood, with storm winds preventing the ebb in between, which is frequently observed to be the pattern during these calamities.

THE HIGHEST TIDE IN THE WORLD

BURNTCOAT HEAD, NOVA SCOTIA

	HW	LW	HW	LW
September 20, 2013	00:47	07:08	13:10	19:31
	14.9 m	0.1 m	14.9 m	0.1 m

I have come to the Bay of Fundy to see the bore, but also because it is the place with the highest tides in the world. It is the most obvious proof that places are not defined only by their solid topography. The idea that the land may be mapped, but that the sea is unknowable because it is unfixed, is a very old one. It persists even on contemporary maps that show every road and river on land, but present the sea in a bland blue wash as if it could not possibly contain any features of interest.

Yet the sea has places too. They may not be fixed on maps or permanent in the sense that printed references demand. Rather, they assume their distinctive character from time to time, predictably to some degree, but also unpredictably, according to the whim of the moon and sun, wind and weather. The best-known sea places are those that are easily seen on the coasts, such as the sand flats of Morecambe Bay. Others lie just off the shore, like the tidal races around Portland Bill and other promontories that troubled Hilaire Belloc in the *Nona*. Occasionally, they may be far out in the ocean.

Tidal bores make for some of the most dramatic of these occasional places. Once, there were rather more of them, but the bores on rivers such as the Seine and the Colorado have been reduced to insignificance by engineering schemes for the benefit of commercial navigation or by siltation and drainage projects. Some, however, are still forces to be reckoned with. The Qiantang River is a commercially important river in eastern China, the artery for the major city of Hangzhou. It, too, has been subject to grand engineering schemes—for irrigation, and later for power—for more than a millennium. But its bore—the world's largest—has not been tamed. It has been an attraction for hundreds of years.

The local myth of the bore is that it is the spirit of a fifth-century-BCE general who was murdered and thrown into the river. Why he revisits the river in phase with the moon is not clear, but the regularity of the tidal surges meant that a table of bore times was issued as early as 1056—making it, in a sense, the first tide table, beating the abbots at London Bridge by a couple of centuries. When the British navy sent HMS *Rambler* to survey the estuary in 1888, captain W. U. Moore was confident there would be no great effect, because the channel was so broad, but the cutters he sent out to make soundings soon found themselves in difficulties similar to those experienced much earlier by Alexander the Great's fleet, with their vessels grounding and crashing into each other in the swell as they shot upriver on an 8-knot current. The water rose 9 feet in ten minutes. Rather than follow the perhaps not entirely well meant advice of local junkmen, the British sailors should have looked at the way the locals secured their boats—on specially constructed elevated platforms far above the usual river level.

All bores must be pumped by a nearby sea that has a large tidal range. The vast Qiantang estuary, where the range between high and low tide is up to 10 meters, functions like the Bristol Channel and the Bay of Fundy in this respect. But why, if the moon and sun exert their gravitational attraction blindly on all the oceans of the world, do gulfs such as these have higher tides than the seas around them?

Here, a new visual image of the tides will help. We must now abandon the astronomers' theoretical ideal of an evenly flooded earth with its equatorial tidal bulges, and instead think about the constraints imposed on real oceans. Within these bounds, it is possible to think of the tide as a giant wave, as the drowning Pincher Martin dimly recalled. When the tide falls in one place, it must rise in another, and so the distance between the two places can be correctly pictured as part of a wave containing both a peak and a trough. This wave will be partially reflected when it strikes the coast or the underwater coastal shelf. If this reflection happens to be returned at the same time as the new tide is coming in, then the two tide waves are superimposed. This is what happens in the Bay of Fundy, which forms the eastern end of the Gulf of Maine. In the western end of the gulf, more open to the ocean, in places like Provincetown and Plymouth and Boston, the tidal range is about 4 meters. But as soon as the tide begins to be funneled into the Bay of Fundy, the range increases—to 6 meters at the entrance of the bay, and then to 10 meters where the bay forks into two further tapering bays. In the southern fork, the Minas Channel, which then squeezes past Cape Split and opens into the Minas Basin, the tidal range reaches an astonishing 15 meters.

These places are "like organ pipes," explains Kevin Horsburgh at the National Oceanography Centre in Liverpool, when I tell him of my plan to visit. "They are a quarter of a tidal wavelength." Kevin talks fast, and draws as he speaks, sketching out a section of a pipe with a quarter wave oscillating inside it. "They respond beautifully to the tidal forcing." Any object has a natural frequency—a frequency at which it will naturally vibrate if stimulated by an external force. A tree waving in the breeze will do so at its natural frequency whatever the wind strength, for example. The familiar analogy of an opera singer's voice breaking a wineglass across a room is even more appropriate. The glass, too, is the shape of a quarter wave. If the bowl of the glass is 10 centimeters high, for example, then a tone with a wavelength of 40 centimeters, equivalent to a frequency of 855 cycles per second, or close to A above the A usually given for an orchestra to tune by, will cause it to resonate. It happens that the natural period of oscillation of the Gulf of Maine is 13.3 hours, very close to the 12.4 hours of the main lunar period. The water here is, in effect, locked into a system that is continually pumped by the moon to reach anomalous highs and lows, like the wineglass made to vibrate so much that it ultimately shatters.

THE BUMPER STICKERS tell me Nova Scotia is "Canada's ocean playground," but the first thing I notice about the Bay of Fundy when I arrive on the shore is that there are no boats on it. The day is fine, the sea is sapphire blue, ruffled by a gentle breeze, but I see not a single sail or even a cargo ship on the horizon.

I'm heading for Burntcoat Head, halfway along the south shore of the Minas Basin, which has the claim to the highest tides of all. The title is contested. Truro, further up the bay, may have recorded higher highs, but it has a smaller range overall because the channel dries to mud and all the water is gone well before the time of low tide is reached. Ungava Bay in the remote north of Quebec Province may produce an occasional higher peak, but Burntcoat Head is the consistent performer on every spring tide. To settle the spirited rivalry that has sprung up between the locations, scientists from the Canadian Hydrographic Service took the first detailed readings using modern instruments at both locations. In 2005, they declared an honorable draw within the limits of measurement accuracy, while neatly underscoring the national claim to distinction: "Both sites have measured tides larger than anywhere in the world—by far," they wrote.

Oceanographers have devised a special metric unit of flow volume that is convenient for describing the most massive sea currents.

It is the sverdrup. Like the newton, the unit of force, and the ampere, the unit of electric current, this unit is named after a scientist, one Harald Sverdrup, a Norwegian who was the chief scientist on Roald Amundsen's North Polar expedition before becoming the director of the Scripps in La Jolla, California, one of the world's most august oceanographic institutes.

One sverdrup is equivalent to 1 million cubic meters per second. One cubic meter of water weighs about a metric ton—almost exactly a metric ton if it is freshwater, a few percent more for seawater, owing to its dissolved-salt content. A million cubic meters would fill a giant cube with sides 100 meters long—the equivalent, in that ever-popular unit of volume, of four hundred Olympic-size swimming pools. The Gulf Stream transports water at a rate of 150 sverdrups where it is at its greatest, some hundreds of miles off the coast of Newfoundland—a figure comparable with other global ocean currents.

In the Bay of Fundy, 160 billion cubic meters of water is shifted during the course of a single tidal cycle. This means that 80 billion cubic meters of water flows in (or out) during the six hours or so when the tide rises (or falls). This works out to an average 4 million cubic meters per second—4 sverdrups. For comparison, the rate of outflow from all the world's rivers sums to just 1 sverdrup. The shifting weight of all this water is even enough to slightly rock the Nova Scotia and New Brunswick landmass back and forth with each tidal cycle.

As I drive around the bay, I notice that the color of the water changes hour by hour. It is not just the changing light. As the flood gathers pace, the bay fills and the silt and sand on the bottom are lifted into the water. The geology is a mixture of volcanic basalt and soft, red sandstone. The former appears in dramatic rocky outcrops here and there, while the latter colors the soil in the fields, the beaches, and the sea itself. The rate of tidal flow is a strong influence. The ability of the water to scour the seabed increases according to the square of its velocity. A tide that runs twice as fast will therefore have four times the power of abrasion. More significant still, the propensity for silt to be carried along in the water without settling increases as the sixth power of the water velocity. For a doubled tidal flow, sixty-four times as much sediment can be carried along. This is dramatically apparent in the bay, where, although the sky is blue, the sea appears brown, red, and purple. Perhaps Homer's famous "wine-dark sea" is not a metaphor or a mistranslation, as is generally thought, but an accurate description of some equally turbid patch of the Mediterranean.

The powerful tidal currents also lift nutrients into the water, making this area important for many marine animals. Zooplankton swept into the bay and caught in a tidal gyre maintains a rich food supply for many species. This abundance attracts large numbers of right whales, which use the bay as a nursery. Small invertebrates may be suspended in the constantly moving water for much of the time, while vertebrates use the tide to hitch a ride, with flatfish species simply lifting themselves off the bottom when the water begins to flow in the right direction. The low tide exposes large areas of the ocean bed, revealing a sediment rich in mud shrimp for hungry wading birds. The shrimp occur in dense populations—ten thousand or more per square meter of mud. A semipalmated sandpiper can consume thousands of the creatures, advancing as the tide recedes with ceaseless dipping of its head, almost doubling its body weight over the course of a single tidal cycle, and thereby obtaining sufficient fuel for its onward migration to South America.

At Burntcoat Head, I squeeze through a narrow cleft in the trees and down some wooden steps onto the flat rocks at the water's edge. It is the hour of the highest tide of the year. A little way offshore is a small island—a low, overhanging cliff of the same red sandstone topped with pines and birch trees. A small gaggle of people has come to look at the tide, but I am not local and I cannot tell whether what I am seeing is remarkable. There is no debris on the tide line, no evidence that any land is flooded—in fact, no sign that anything out of the ordinary has occurred. I have to remind myself that what I am looking at is not a freak event, but a relatively normal event in a freak area. Even though this is the highest tide of the year, a very high tide comes here twice a day. There is none of the usual flotsam I would see on a high spring tide in Norfolk—reed stalks and feathers and crab shells lifted off the marshes—because any marshes have long since been washed away entirely by the relentless power of the tides. Only bare rock remains, and even that is being constantly eroded.

I return late in the afternoon when the tide has gone out. Nobody has come back to see this, but it is in many ways a more interesting spectacle. The island is no longer an island and is now like a battleship left high and dry on the rocks. I walk out on the seabed. The rocks are mainly red sandstone, but here and there are boulders and pebbles of a harder greenish stone that make a painterly contrast. It is like the surface of another planet. About a dozen meters "down," I pick up a lost fisherman's weight. I classify it as lagan. Blessedly, it is the only human debris I see. But then, most of our rubbish floats, and will have been swept out to sea by the ebb tide.

I notice three kinds of seaweed: a delicate wrack; another dark, frondose species; and one with curled leaves like lollo rosso lettuce. In the deeper places, patches of sea grass wave gently to and fro in the breeze as they will do again underwater in a few hours. It feels like an ordinary low-tide beach until I look back to the shore and see how far out and how far down below its level I have come. An idle diversion has become a transgression. I am in an imaginary world, a transient landscape never inhabited, never mapped, never captured by a painter. I am alone, yet people have come here at low tide before. Here and there along the intertidal cliffs, I find that pairs of lovers have carved their names in the giving surface. I wonder how long they will last.

ART AND ENGINEERING

When he came to the Nova Scotia coast to make new work in the winter of 1999, the environmental sculptor Andy Goldsworthy says he felt at first out of touch with the place because of the unfamiliar cold and because of the extraordinary tides. His habitual solution to get to know the place, the waters and the land, was to begin to make artworks using the local materials—stone, ice, driftwood. Often these were in the intertidal zone where they would soon be altered or destroyed, and their ingredients reclaimed, swiftly or slowly, by nature. As the works were being made, the next tide was always the absent force that would have to be felt before they were truly complete.

The landscape became Goldsworthy's palette when he was still an art student at Preston Polytechnic in Lancashire, where it served as a useful corrective to the self-absorbed business of academic pedagogy and making art about art. "Working between the tides is among the very first work I made outside," Andy tells me. "It was on the beach at Morecambe. In Lancashire, I was in a studio, but every evening I saw this big expanse of beach, and wanted to work on it. I dug holes and made lines, and then watched the tide coming in and washing those lines away. The tide represented this intensity of a workplace where you weren't in control. There was a huge sense of resistance in the place. The tide was a real teacher."

For Andy, the tide is one of nature's artists, a maker of places. "And it changes that place radically each day, so it's a charged process. When you work on the same beach time and time again, the material of it changes, things appear and then get covered again. And there's the rhythm of each tide—wonderful calm days when it just slides in, and then when it rages and comes in great storm waves."

One of the works that Goldsworthy made on the rocky Fundy shore was a large, nest-like structure of loose driftwood shaped rather like an igloo. It floated off whole as the tide rose up the beach where it had been assembled, gradually shedding its timbers one by one. Eddies in the fast-rising water seemed to sense the circularity of the sculpture, catching it in a kind of dance. The very force that breathes life into the work will also be the thing that destroys it, the artist noted in a film about his work with water. Goldsworthy also built some large, egg-shaped cairns from stones he found on the beach. One cairn was positioned in a place where it would be completely covered when the tide came in. Its disappearance so soon after its making was to be an essential part of its story. Perhaps it would survive that first tidal cycle; perhaps it would survive several. But what was certain was that it would not last. The natural world would prevail. In the film, time-lapse footage of the tide rising over the beach cairn was intercut with footage of a similar cairn in a field. Ferns grow up around this construction, and in one segment of the film, a Highland bull scratches himself indulgently against it, a vital emblem of the power of nature. "I haven't simply made the piece to be destroyed by the sea," Goldsworthy explained of the cairn on the beach. "The work has been given to the sea as a gift, and the sea has taken the work and made more of it than I could ever have hoped for." The sea accepts its gift with gentle hands.

The film is captivating, but I realize there is also something deceiving about the work it shows. As we watch them being steadily immersed or teased apart by the waves, these artworks radiate a sense of relaxed submission to their fate. Yet I have become accustomed to dashing here and there in order to meet my rendezvous with the tide. What intrigues me about Goldsworthy's work on the foreshore is not only its ephemerality—many artists seek to undermine the art world's demand for permanence by creating work that in one way or another will not last—but its demanding instantaneity, the unshown and unspoken haste with which it must be put together, rather than the manner of its destruction. Perhaps it is something to do with the way the filmmaker has chosen to document the artist at work, but I feel I am missing a sense of the urgency that was surely present as he assembled his work in the short interlude between the tides. These are monuments that could stand for days or centuries, yet they have had to be made in just a few hours. "That underlying pressure is huge," Andy admits. "It is somewhat frantic work. The pressure was obviously there, but the best way to get the job done is to do it quickly, quietly,

and efficiently, and to work with the rhythm of the time you've got."
Not everybody appreciates this, it seems. On another occasion, Andy
was working on the low-tide beach in Guernsey, transforming the flat
beach into a surreal landscape of rounded sand boulders with a group
of student helpers. At one point, he looked up from his task to find
they had all stopped for a game of volleyball. "Then I went ballistic,"
he says. "I said some things I should not have said."

IT IS NOT ONLY ARTISTS and whales that are drawn to the Bay
of Fundy by the power of the tide. The bay is also one of the prime
locations for testing the practicality of schemes to convert this power
into electricity. It has been estimated that the whole bay has a theoretical power output of 6,000 megawatts, the equivalent of a few large
nuclear power stations. Although only a tiny proportion of it could
ever be economically extracted, the total power dissipated by the
action of the world's tides is more than 3 terawatts (3 million megawatts), which is three times the global human consumption.

Before I leave the bay, I pay a visit to two experimental tidal-power
projects. The first was instigated in the 1970s in response to the global
oil crisis and is now ancient history in the terms of this youthful
industry. The installation, at Annapolis Royal on the south side of
the Minas Basin, is housed in a shiny aluminum control tower and is
designed to generate up to 20 megawatts by controlled release of the
head of ebb water built up by timed damming of the tidal Annapolis
River. On the other side of the basin, at the end of a long stone-chip
road through the forest not far from Parrsboro, is the much newer
wood-and-glass visitor center of the Fundy Ocean Research Center
for Energy (FORCE), a consortium set up to provide facilities where
engineering companies from around the world can test "in-stream"
devices for extracting tidal energy in the most challenging, as well as
the most promising, of environments. Their devices would lie on the
open seabed rather than requiring the large-scale coastal intervention

of dam-based schemes. Colorful cutaway diagrams present a bewildering array of experimental proposals—some like giant ships' propellers, others based on turbines resembling jet engines, others more like paddles. The technology is at a stage where consensus has yet to be reached even on the basics of what constitutes the best design. But ideas for wind turbines were once just as varied. Suitable turbines on the seabed stand to harvest energy that is much more concentrated than wind simply because water is eight hundred times denser than air.

But the prize is obvious to all. Everywhere I have been able to look out to sea, I have seen clear blue waters churned to white, ripping past rugged headlands. Where there is no rocky obstacle, broad, glossy streams slide rapidly along, creating busy overfalls—lines of small breakers—where they run over patches of slack water. The sheer speed of these currents explains why I have seen no boats out in the "ocean playground."

At the FORCE center, I am looking out at a black rock in the bay near where the test rigs are to be sunk, when I am joined by the manager, Mary McPhee. The tide is flooding fast, and the difference in height of the water from one side of the rock to the other is quite apparent; it can be as much as a story of a building, Mary tells me. There are no big waves, just lots of surface agitation, trails of whirlpools, and patches of energetic ripples. "It has a very riverine acoustic here," Mary suggests.

Both of these well-organized visitor centers leave me with the impression that winning public support for tidal power is just as important as conquering its formidable technological challenges. But their different eras also offer a hint that the journey has not been easy, and perhaps will not be easy in the future either. The Annapolis scheme was the product of a time of simpler technological optimism. The small plant was once intended as the forerunner of a barrage that would have been placed across the narrows of the Minas Basin, creating a tidal dam of vastly greater power-generating potential. (A similar barrage across the Severn estuary was also seriously evaluated at this time.) But no turbines could be found that were suited to the conditions of the bay. Now the technology is more advanced, but concerns have grown too. Tidal power is "green" in the sense that it is carbon-neutral in operation, and the latest schemes are likely to make a relatively small visual impact in the landscape, since the generators will be out of sight on the seabed, so they should not raise the objections that some land-based wind farms have done. But there are now additional environmental concerns to be addressed that were not recognized a few decades ago. Mary is keen to emphasize the monitoring that is being done, concurrent with tests on the turbines themselves, to assess the likely effect on populations of sturgeon and the sixty other species of fish in the bay.

Perhaps there is a generational shift in tone. When I ask Les Smith, the guide at the Annapolis site, about the future prospects for tidal-power generation in the Bay of Fundy, he is notably less sanguine. "It has no future. It's really not suited for such a heavily sedimented body of water." But Mary is resolutely upbeat. When I ask her which of the trial technologies she'd bet on, she shoots back, "I think they're all going to work." Eventually, they may do—although perhaps somewhere else with cleaner waters, such as the Pentland Firth or elsewhere in the Orkneys where there are also schemes undergoing trials. In the Bay of Fundy, the tide wrecked the first turbines that were tried in the space of thirty-six hours. Theories vary as to why. Perhaps excessively turbulent fluctuations in the flow were responsible, or perhaps the sheer mass of waterborne sediment simply proved too much for the mechanism to handle. Although vast amounts of energy are bound up in the tides, it is still a small total compared to the energy theoretically available from geothermal and solar sources, both of which raise fewer technical difficulties. To be sure, tidal power will only ever make up a small part of our energy budget.

A VICTORIAN OMNISCIENTIST

The high tides in the Bay of Fundy are accentuated by the wave that is raised because of the shape of the coastline. The Gulf of Maine, of which the bay is part, bends round to form what is effectively a closed basin, which leaves it largely independent of the tidal movements of the Atlantic Ocean. The resonance effect in the gulf is highly sensitive and leads to some curious side effects. Phillip MacAulay, head of tides at the Canadian Hydrographic Service Maritimes region in Halifax, Nova Scotia, explains to me that even the small Annapolis barrage has altered the tides in the Bay of Fundy, but it has also produced a change in the tides at the far end of the gulf, in Boston. The variation has been measured to be about 3 centimeters. Had the larger scheme for a barrage across the whole Minas Basin gone ahead, there is a chance that a far larger variation would have been seen, perhaps placing Boston's low-lying Logan Airport at significant risk of tidal flooding. Even more astounding is the likelihood that a barrage across the Minas Basin would alter the tidal movements in the Severn estuary, more than 4,000 kilometers across the Atlantic, not in this small way, but possibly producing very large changes in the tides locally in a dramatic example of the opera singer–wineglass effect. Of course, a future Severn barrage might alter the Fundy tides too, because of the similar resonant frequencies of the two bodies of water.

The idea of tide as a long wave is useful in visualizing the anomalous goings-on in places like the Bay of Fundy. But it can also be generalized to provide a new way of looking at ordinary tides anywhere at sea. The wave in the bay is trapped; it is what physicists call a standing wave. The tidal waves in the ocean are essentially standing waves too, but they appear to travel because of the effect of the earth's rotation. Bede long ago sensed that the rising tide swept south along the English east coast from his monastery on the River Tyne and around the bulging coast of East Anglia to the Thames in just such a wave.

To see how such waves arise, imagine once again that the planet is uniformly flooded with seawater. If there were no continents, and if there was no friction between the sea and the seabed, the tidal bulges in the oceans raised by the moon and sun would slip around the planet in a westerly direction at a speed of nearly 1,700 kilometers per hour, measured at the equator. The tidal range would nowhere be more than about half a meter. In reality, we can see that the progress of the tide is impeded by the north–south barriers thrown up by the continents. In addition, waves travel at speeds governed by the depth of the medium they travel through, and the oceans are too shallow for the tidal bulges to keep up in their chase of the moon. Furthermore, even the open ocean is composed of more or less distinct basins with their own natural frequencies of oscillation, which become important in knowing the local tides.

All these influences mean that high and low tides, though spaced apart at the *interval* dictated by the moon's orbit, seldom fall at the *time* they would be expected, when the moon is most directly over-head. Misled by this offset, Galileo had assumed that the observed period of the tide was, in fact, independent of the lunar period, leaving it to Laplace to prove the connection much later. Not only is the timing of the tide thrown off, but so are the heights that we might expect to see. In many places, as we know, the tidal range is far greater than half a meter, though in few places is it so exaggerated as it becomes in the Bay of Fundy or the Severn estuary. In one or two other places, it even happens that the oscillation due to the basin cancels out that global wave, and there is hardly any tide.

Evidence that the tide travels like a wave came not from astron-omers and mathematicians like Newton and Laplace, but from the growing volume of empirical data being gathered by naval expedi-tions to remote parts of the world in the name of empire and trade, as well as scientific discovery. Ports on many more coasts were now in a position to supply direct measurements of the tides. Even remote oceanic islands were included. The Royal Navy explored far and wide, even Charles Darwin's ship, HMS *Beagle*, contributing its share of tide measurements. The French joined the project. The independent United States under Thomas Jefferson, who still surely ranks as that country's most scientific president, instituted a survey of its own long coastline. The first aim was to produce serviceable charts of American ports, but the task was soon extended to making detailed tidal observations. One remarkable scientific collaboration in June 1835, involving Britain, the United States, and seven European countries with an Atlantic seaboard from Norway to Portugal, coordinated measurements taken every fifteen minutes over a period of twenty days from 650 tidal stations around the world.

In many ways, it was the forerunner of the "big data" studies that have become voguish since the advent of powerful supercomputers. The man who, above all others, took it upon himself to see whether these data could be collated in such a way as to tell us something fundamental about the tides was the Reverend Professor William Whewell. Whewell was a polymath of astonishing breadth of knowledge, who escaped following his father into trade as a carpenter because of his mathematical ability, which led him ultimately to the Master's Lodge at Trinity College, Cambridge. The author and wit Sydney Smith said of him, "Science is his forte, and omniscience his foible." Whewell indeed popularized the use of the word "scientist" and introduced the natural sciences tripos at Cambridge University. He died in 1866 of injuries sustained in a riding accident, at the age of seventy-two.

Whewell also invented the horrible word "tidology" to describe his project of nearly twenty years to make sense of the new measurements. In this enterprise, he enjoyed the support of the long-serving naval hydrographer Francis Beaufort (after whom the Beaufort scale of wind strength is named), under whose aegis more than a thousand Admiralty charts were first issued. Using the tidal data from so many sources, Whewell was able to draw his own maps of the world's oceans, cut across by what he called "cotidal lines"—lines joining locations where the tide was high at the same time, like weather map isobars. These lines gave a graphic impression of the tide as a single majestic wave, sweeping progressively from New Zealand through the Southern Ocean and then deflecting through the South Atlantic and on up through the North Atlantic toward the Arctic. Tide readings from the Pacific coast of the Americas hinted at a similar wave moving south, although there was insufficient data to complete the picture for the entire ocean. Whewell's visualizations were compelling new evidence of how the tides really move around the earth, well judged to appeal to an imperial establishment more comfortable with maps than with equations.

The picture wasn't quite as straightforward as a single wave, of course. Places such as the Bay of Fundy and the Severn estuary were already recognized as exceptional. Similar anomalies were recorded on the shallow Argentine shelf where the Atlantic tide wave is augmented by waters flowing through the Drake Passage from the Pacific to produce some places with tides nearly as great as those in the Bay of Fundy. Further up the same coast, the *Beagle*'s captain, Robert Fitz-Roy, found a spot in the Plata estuary where there seemed to be no tide at all. Giving the Bakerian Lecture at the Royal Society in 1847, Whewell revealed that "there are some places where the tides are regulated by the sun as much as they are by the moon, or even more." At Tahiti, for example, the time of high tide is never far from noon, and the tides are very small, ranging little more than half a meter. Indeed, this may be one reason why the inhabitants of such island groups were able to develop an interisland canoe culture for trade. "The tides over a great portion of the central part of the Pacific are so small, that we may consider the lunar tide as almost vanishing," Whewell concluded. These local features were later termed amphidromic systems, in an obscure reference to the ancient Greek custom of naming a child after first carrying it around the hearth, because the tide wave circulates around these still points.

It was not necessary to voyage into exotic waters in order to discover such tidal oddities. Domestic waters could be just as puzzling as the South Seas. In particular, Whewell's cotidal lines for the North Sea progressed smoothly down the English coast, as Bede had foreseen, but readings from the Continental coast did not follow suit. Somewhere in the middle, the lines had to stop. In 1837, Captain William Hewett led the survey vessel HMS *Fairy* in an attempt to locate the amphidromic system suspected to lie somewhere out in the southern part of the North Sea. Hewett was highly knowledgeable about the coastal waters of East Anglia and had made many surveys in the area, especially around the ports of Great Yarmouth and Lowestoft. Shortly before setting off, he had written a paper in *Nautical Magazine* alerting sailors to recent shifts in the sands off Yarmouth. This part of the Norfolk coast was known for its long stretch of good ground protected by sandbanks offshore where hundreds of ships could lie safely at anchor. It was where the North Sea fleet would rendezvous in time of war. But Hewett had intelligence that the sandbanks were shifting, altering the line of entry into this safe channel, and he now warned that the protective banks might one day be washed away altogether, as the Scroby sands had been long before; they were once a "green isle," he wrote. (A stable island emerged from the sea in 1578, and Yarmouth people adopted it for fairs and picnics at low tide; in 1582, a local landowner tried to claim it for himself, and it was promptly washed away by righteous storm waves.)

In the summer of 1837 and again in 1840, though, Hewett's mission, undertaken on the orders of the hydrographer Beaufort and at Whewell's request, was to station his vessel repeatedly at fixed positions in this dreary patch of water and take depth soundings by lead line over the full tidal cycle in order to establish the variation in the level of the sea. By the end, Hewett had identified not one, but two amphidromic points where there is no tide—one directly between East Anglia and the Netherlands, and another west of Jutland.

Soon after this work was complete, in November 1840, Hewett's ship was caught in a severe gale and wrecked off the coast he knew so well, with sixty-three men lost. No convincing wreckage was ever found, and the whereabouts of the wreck are still not known. After hope had finally been given up, *Nautical Magazine* published these lines: "Although the loss of such an accomplished officer as Capt. Hewett is irretrievable, yet, we feel great satisfaction in announcing to the maritime world, that the major part of his noble survey of the North Sea is on the copper, and will be published forthwith."

9

IN GREAT WATERS

THE MORAL TIDE

The music that closes Benjamin Britten's great debut opera, *Peter Grimes*, washes in gently at first: high strings suggest the flat, shining sea of the English east coast at dawn. Woodwinds make the first ripples, and the swell deepens with the brass, before the cymbals bring the first waves crashing together, making spray that scintillates in the low sunlight.

As the orchestral sound intensifies, the chorus joins in:

> With ceaseless motion comes and goes the tide,
> Flowing, it fills the channel vast and wide;
> Then back to sea, with strong majestic sweep
> It rolls, in ebb yet terrible and deep

The lines are not, in fact, an ending. They come from near the beginning of George Crabbe's long poem *The Borough*, which is arranged in twenty-four "letters," each describing a facet of life or a personage of a coastal town, which Crabbe based in part on Aldeburgh in Suffolk, where he was born and was briefly a curate.

Aldeburgh is a town much ravaged by the sea. The sixteenth-century town hall, once the central feature of a marketplace that lay two blocks inland, now stands exposed and alone on the seafront, and Crabbe's own birthplace on the shingle spit known as Slaughden that runs south from the town, squeezed between the North Sea and the serpentine River Alde, is no more, swept away in the great storm of 1953.

In the 1760s, the young Crabbe would accompany his father on the fishing boat in which he owned an interest. This direct experience of the sea gives his verse a rare marine authenticity. Though many poets romanticize the sea, Crabbe is one of few who really understand its looks and actions, and *The Borough* is alive to every state of the tide and the town's dependence upon it. His next stanza accurately describes the ecological banding of vegetation along the tide line:

Here Samphire-banks and Saltwort bound the flood,
There stakes and sea-weeds withering on the mud;
And higher up, a ridge of all things base,
Which some strong tide has roll'd upon the place.

Crabbe uses images of the tidal flood and ebb as a moral warning. So futile are the borough churchmen's efforts at preventing sin among the townsfolk that they might as well be directed to stemming the tide.

Although Grimes's story is the best known in the set, thanks to Britten's opera, the most terrible scene comes in Letter IX, "Amusements." A party is rowed out to a shingle bank exposed at low tide, where they are able to enjoy a vicarious experience of the great expanse of the sea: "The watery waste, a prospect wild and new." Such banks are a feature of the shallow waters off East Anglia, built up in lines parallel to the coast from sediments of sand or gravel ceaselessly scoured from nearby beaches and cliffs—half a million cubic meters of material annually finds itself redistributed in this way to make the banks off Great Yarmouth, for example. They are not entirely stationary features, but liable to creep treacherously along the seabed, as the ill-fated Captain William Hewett was well aware, although the exact means by which they are formed and shifted around are still not well understood.

Debarked onto the bank, the party enjoys singing and dancing and hot refreshments. A few go beachcombing. "And not a grave or thoughtful face was found." Meanwhile, the sailor and his mate who took them out indulge in stronger drink and fall asleep, and their boat drifts off on the rising tide. A "lady sage" notices, and raises the alarm. Suddenly, they all understand the importance of the tide to which before they had been so inattentive. They shout for help from the land but are not heard:

> They shout once more, and then they turn aside,
> To see how quickly flow'd the coming tide;
> Between each cry they find the waters steal
> On their strange prison, and new horrors feel;
> Foot after foot on the contracted ground
> The billows fall, and dreadful is the sound;
> Less and yet less the sinking isle became,
> And there was wailing, weeping, wrath and blame.

Miraculously, a crew of more competent seamen spies the empty, drifting boat and comes to rescue the party, all thoughts of godless amusement now banished from their minds.

The tide asserts its power late in the tale of Grimes too, whose story is offered as a sketch of one of the poor of the borough (Letter XXII). Grimes, deserted by his own son, takes on an apprentice boy, who is later found dead. Two more such boys die in mysterious circumstances at sea, and now the townspeople will not allow him to take another. Forced to fish alone, his listless existence is regulated only by the slow beat of the tide:

> Thus by himself compell'd to live each day,
> To wait for certain hours the tide's delay;

At the same times the same dull views to see,
The bounding marsh-bank and the blighted tree;
The water only, when the tides were high,
When low, the mud half-cover'd and half-dry;

Grimes is eventually run out of town and driven to madness. At length, to the few who can muster any sympathy for him, he recounts fragments of what happened to the boys. His confused tale is mingled with repeated visions of his father, who appears as an angry spirit carried on the flood tide with the boys' ghosts on either hand.

Crabbe's poem makes it clear that Grimes was violently abusive toward the boys. Britten is more ambiguous toward the fisherman, and more critical of the vindictive townsfolk, who finally instruct Grimes to take his boat and sail out of sight of land and then sink her. By the closing chorus of the opera, it is the next morning. There is a report of a boat sinking out at sea, but nobody is interested. The life of the town goes on. And with it the tide.

LOVE AND LOSS

When did the tide come to be imagined as a moral force as well as a physical one? Why should a blameless cycle of nature be chosen as a metaphor of punishment and mortality? The sun and moon do not judge us in this way. Is high tide good and low tide bad in this scheme, or is it the other way around? Is the ebb to be preferred to the flood? How does this notion really work?

It must start with the fact that, unlike celestial bodies, the tide has the power to take human lives. Who does it claim? It claims the ignorant, the reckless, and the merely unlucky. But the sea would claim many of these lives even if it did not rise and fall, so we must look further if we are to explain why the tide specifically carries the symbolism that it does.

The tide introduces the vertical movement of the sea. It is surely significant that we naturally tend to speak of the moods and emotions that govern our moral behavior in terms of high and low. In this, the tide joins many natural metaphors used to describe high and low mood. High tide represents hope and opportunity—the "tide in the affairs of men, / Which, taken at the flood, leads on to fortune." Low tide stands for the loss of these things, as we see in John Betjeman's poem "Youth and Age on Beaulieu River." The poet's tanned erotic fancy here is not the tennis-playing Miss Joan Hunter Dunn but "Clemency, the General's daughter," who sails out on the tide and "Will return upon the flood," while old Mrs. Fairclough eyes her enviously through binoculars: "the older woman only / Knows the ebb-tide leaves her lonely / With the shining fields of mud."

This vertical movement, unlike that of the poet's skylark or the burrowing mole, say, also occurs within a certain range, within what one might, in fact, be tempted to call an *appointed* range, its high and low limits apparently set by some godlike power. This aspect of the tide is reflected in our adjective "tidy," which derives from the Middle English "tide" and first meant "seasonable"—hence, later "neat," as of something "in its place."

This sense of the tide operating within defined bounds is clear in the lines of the 1860 hymn known as the "Navy Hymn":

Eternal father, strong to save,
Whose arm doth bind the restless wave,
Who bidd'st the mighty ocean deep
Its own appointed limits keep:

The hymnist William Whiting based the work on Psalm 107, in which storm waves terrify "They that go down to the sea in ships, that do business in great waters." This being the Mediterranean, however, there is no mention of the tide; the suggestion of tidal movement was the addition of the British hymnist. In a powerful scene in another of

Britten's operas, *Noye's Fludde*, the audience joins in singing the hymn as the flood rises and Noah's Ark is floated on the stage.

The tidal metaphor provides a useful safety valve for our emotions because we know that its vertical movement is both regular and cyclical. It can be a comfort when we feel low, because we know that high tide will come. It provides a salutary check when we are in high spirits, reminding us that euphoria cannot last.

But the tide is a dynamic force. It is not only the extremes of high and low that characterize its role in our lives, but the transitions between the two, the flood and the ebb. Depending on where we are standing, the flooding tide is a threat, as Crabbe so powerfully demonstrates. Any nineteenth-century reader of his poem would be instantly put in mind of the biblical flood, which was God's punishment for humanity's evil. The rising tide, then, provides a twelve-hourly reminder of our sinfulness. Yet the flood is also life-giving. Water generates life of all sorts and is the symbol of that life. The idea that animal life emerged from the water is contained in Genesis and the Qur'an, as well as in canonical evolutionary theory. And of course, mammalian birth is accompanied by a flood of amniotic fluid.

The ebb is, in many ways, more interesting psychologically. The belief that no creature can die except when the tide is ebbing is attributed to Aristotle, and has been repeated, if not necessarily endorsed, by many distinguished writers since, from Pliny the Elder to Sir Thomas Browne and James Frazer in *The Golden Bough*, as well as Dickens. The belief is perhaps the inevitable corollary of the less arguable association between birth and the flood.

This piece of folklore has doubtless also been perpetuated by the custom, in seaside parishes, of recording parishioners' time of death in relation to the state of the tide. In Shakespeare's *Henry V*, Mistress Quickly tells how Falstaff died "just between twelve and one, ev'n at the turning o' th' tide." However, in the closing chapter of *Moby-Dick*, as Captain Ahab is lowered into a rowboat to do final battle with the White Whale, he tells Starbuck rationally enough, "Some men die at ebb tide; some at low water; some at the full of the flood."

Parish records confirm, of course, that people die at any state of the tide, and the custom of noting the death in relation to the tide should be appreciated more as an indication of the former importance of the tide in shaping coastal communities' sense of time than of anything else. Nevertheless, in some places this belief persisted even into the twentieth century. David Thomson's *People of the Sea* records a Mayo fisherman's thoughts as his wife dies on the ebb tide: "I thought to myself, and I still praying, if God spares her now for these few minutes, and the tide to turn, she will be safe." Of course, God does not spare the tide to turn: linking death with the inexorable tide is a way of accepting its inevitability.

Alfred Tennyson was impressed by the sea throughout his long life. As a boy growing up in Lincolnshire, he was taken for holidays to Mablethorpe and Skegness, where he would recite his poems out on the mudflats at low tide. He especially enjoyed stormy days when he could watch the breakers crashing on the broad sands. He later settled at Farringford near Freshwater Bay on the Isle of Wight, where the island is almost cleaved in two by the vigorous tides. In *Maud*, he wrote of "Listening now to the tide in its broad-flung shipwrecking roar." This is no romantic fancy; the island's exposed south coast was indeed a shore of wrecks and smugglers.

"Crossing the Bar" is Tennyson's famous anticipation of his own death. It was written quickly, in 1889, when Tennyson was eighty years old, during a crossing of the Solent to the Isle of Wight. In the poem, he gives a scrupulous description of the mighty sweep of the tide: "But such a tide as moving seems asleep, / Too full for sound and foam." The flood has borne him far and wide in his life, he records, but now the tide "Turns again home," and it is this swift ebb—calm and silent, yet also powerful and undeniable—that will now carry him away "to see my Pilot face to face / When I have crost the bar."

Three more British seafarers allow themselves to be carried here and there on the flood tide in Robert Louis Stevenson's 1894 adventure *The Ebb-Tide*, which was coauthored with his stepson, Lloyd Osbourne. The sailors arrive on an idyllic Pacific island in a stolen ship, the *Farallone*, hearing first the breakers on its reef, before the flood tide obligingly lifts them into a lagoon sheltered by the walls of a coral reef:

> Twice a day the ocean crowded in that narrow entrance and was heaped between these frail walls; twice a day, with the return of the ebb, the mighty surplusage of water must struggle to escape. The hour in which the Farallone came there was the hour of the flood. The sea turned (as with the instinct of the homing pigeon) for the vast receptacle, swept eddying through the gates, was transmuted, as it did so, into a wonder of watery and silken hues, and brimmed into the inland sea beyond.

The three are a dissolute bunch who spend their time squabbling and drinking, having been previously thrown off various ships. On the island, they meet a compatriot, scarcely better than they are, who is amassing a hoard of pearls that he hopes will make him rich if he ever gets home. Why the "ebb-tide"? The men are literal drifters, arriving where they do as a result of being "taken by the flood"; the ebb, though, is a symbol of moral turpitude, the measure of the depths to which the four Britons have sunk.

The tide is given a different moral power in another Victorian poet's best-known work. Matthew Arnold's "Dover Beach" has been much admired and much puzzled over since its publication in 1867, and it is still a popular text for analysis in literature classes. The short poem begins as a simple nocturne: "The sea is calm to-night. / The tide is full, . . ." But the mood quickly darkens. A light on the distant coast goes out; the pebbles on the beach make an incessant "grating roar"; the poet's mind turns to miserable thoughts, and then to this grim aperçu:

> The Sea of Faith
> Was once, too, at the full, and round earth's shore
> Lay like the folds of a bright girdle furl'd.
> But now I only hear
> Its melancholy, long, withdrawing roar,
> Retreating, to the breath
> Of the night-wind, down vast edges drear
> And naked shingles of the world.

This earth, Arnold concludes, "Hath really neither joy, nor love, nor light, / Nor certitude, nor peace, nor help for pain."

It seems at first that the poem might be a lament on the crisis of Christian faith induced by the publication of Charles Darwin's *On the Origin of Species* in 1859. But Arnold is believed to have conceived the lines when staying at Dover on his honeymoon in June 1851. (Whether Mrs. Arnold enjoyed the trip as much as her husband is not recorded.)

But the idea of faith as a tidal sea immediately raises complex questions. Arnold's overt pessimism seems to indicate that faith might withdraw forever, but the metaphor tells us it will return. Perhaps it shows the poet's faith in faith. Yet the supposition that faith may rise and fall, that it is cyclical, subject to some kind of celestial mechanism, is itself surely a profane thought. The poet reconciles himself to something more like classical stoicism than like the stupid reverence of Victorian churchgoing.

The fashion for tidal metaphor during the Victorian period is surely explained in part by growing awareness in a land no longer "hedged in with the main," as Shakespeare has it in *King John*. A maritime empire ensured that the sea was no longer a vast unknown. Its depths were charted by the British Admiralty, its lengths crossed by exotic goods and, increasingly, by leisure travelers. Seaside resorts grew up where people had once only feared the sea and turned their backs on it. Now, they began to look out on it, and to paddle and swim in it. Aldeburgh is one of many old coastal towns where the dwellings show ample evidence of both of these attitudes toward the sea.

Today, however, our employment of the tide as a metaphor has grown rather more casual than the poets'. We forget it is a cyclical phenomenon, in which each station and action is bound to recur. Instead, the flood becomes a one-off cataclysm. News reports warn of a flood of immigrants, of data, of cheap imports. That's perhaps forgivable, since a flood is not only a flood tide, but a one-off pluvial deluge. But the ebb is treated with the same lack of regard and seen as a terminal loss of opportunity. When we say that hopes of finding survivors after an earthquake or similar disaster are "ebbing away," we really mean that hope is lost. If the chances of a successful outcome of a summit meeting are "ebbing away," we mean to say that that chance will not come again. If we were more precise about it, we would be suggesting almost the opposite: that another chance, just as good, will come along pretty soon.

THE GREAT WAVE

The Victorian writer who comes closest to sensing how the tides really move may be Thomas Hardy, who is not usually thought of as a marine author. In *The Return of the Native*, however, he draws a telling contrast between the "ancient permanence" of Egdon Heath and the restless sea that all but surrounds this area of wilderness reaching out along the Purbeck peninsula in Dorset: "Who can say of a particular sea that it is old? Distilled by the sun, kneaded by the moon, it is renewed in a year, in a day, or in an hour."

For the sea is indeed molded over periods long, short, and intermediate, as the scientists of this time were graphically beginning to show. To see how this is so, it is useful to return to the model of the tide as a single wave, explored by William Whewell. The idea of the tide as a giant wave roving the planet's oceans seems like a hangover from Norse mythology, but it turns out to provide a powerful new way of visualizing its numerous gravitational component causes.

What kind of wave is it? Mathematicians had known for some time that any wave, no matter how complicated or unnatural its shape, can be broken down into constituent "pure" waves that may be described by certain simple equations. These constituent waves, of various lengths (periods), heights (amplitudes), and lateral displacements (phases), can be scaled appropriately and then added together to reconstruct the original wave in a process known as harmonic analysis. The convenience of this method for physicists is that it makes it possible to break down the energy of any wave into these distinct harmonic constituents, which may be handled mathematically in ways that the complex original cannot be.

As the word "harmonic" implies, this way of thinking about waves is informed by the way musical sound is produced. When a musician ostensibly plays a single note on an instrument, what we hear in the audience is not, in fact, one note. It is a combination of the intended note at one frequency, which musicians call the fundamental, and simultaneously emitted notes at multiples of that frequency, which are called overtones. The fundamental is also referred to as the first harmonic, and the overtones are numbered as second harmonic, third harmonic, and so on. The ear does not hear these notes separately, but experiences them together as the timbre, or distinctive sonic quality, of the instrument. The burnished tone of the horn comes from the combination of the fundamental with contributions from the first few overtones. The very different, "reedy" quality of the oboe, on the other hand, arises because of strong contributions from some of the higher harmonics.

The traveling wave of the tide is more complicated than this because it is made up of many fundamentals, as well as their overtones. Each of these fundamentals, or harmonic constituents, has a single gravitational cause. If it helps, you can imagine that each harmonic-constituent wave is caused by a single satellite, exerting just the right pull and orbiting at just the right speed around the earth to account for its amplitude and period. This trick certainly helped Laplace, who took the idea further, imagining that each harmonic constituent was generated by a celestial body, which he collectively called *astres fictifs* ("fictitious stars"). This is true, of course, of the major constituents of the tides raised by the purely circular component of the orbits of the moon about the earth and the earth about the sun. But Laplace's ruse allowed departures from this ideal situation to be thought of in the same way, so that the true elliptical nature of these orbits, as well as other irregularities, could be clearly separated in the analysis. Thus, the elliptical deviations of the moon's and earth's orbits could be conceived as two *astres fictifs*. So could the variation in the moon's position in relation to the earth's equator, the earth's own pull on the waters on its surface, and even the gravitational corrections owing to the fact that the earth is not perfectly spherical or uniform in density. Even the small gravitational pull on the rest of the ocean produced by the mass of water displaced by the tide itself could be accounted for in this way.

In the Bay of Fundy, for example, the tidal range owing to the moon only is about 12 meters. The remaining 4 meters of its range come from other harmonic constituents. In Tahiti, by contrast, the lunar constituent of the tide is practically zero, and the small tide arises almost entirely from the smaller solar constituent. The tides in any part of the world are simply a sum of harmonic constituents. Real measurements of the tide taken over a suitably long period can be deconstructed by harmonic analysis into these pure constituents, which include not only the twice-daily and daily lunar and solar tides, but also monthly lunar and annual solar variations and other factors (and their overtones). The harmonic constituent with the longest period—18.6 years—arises from movement of the plane of the moon's orbit in relation to the earth's equatorial plane. For practical purposes, then, it is only necessary to have nineteen years' worth of tidal measurements in order to publish a fully accurate predictive tide table for any given place.

By the mid-nineteenth century, many places around the world had the necessary records. All that remained was to process the data. Lacking the computers that would make light work of such a project today, the British Association for the Advancement of Science in 1867 established a committee to make a start on the tedious arithmetic. The project was led by William Thomson, later Lord Kelvin, one of the most versatile scientists of the age. Thomson had entered Glasgow University at the age of ten and went on to occupy that institution's chair of natural philosophy for more than half a century. After his death in 1907, he was buried in Westminster Abbey next to Isaac Newton.

Thomson was in many ways the epitome of the Victorian scientist, all-curious, but aware, too, of the duty to apply his work for practical betterment. He was responsible for formulating both the first law of thermodynamics and the law of conservation of energy, which states that heat and work are equivalent, as well as developing the related concept of absolute temperature, which is still measured on the Kelvin scale. These ideas were important for the development of engines and refrigeration, while his expertise in electromagnetism guided the project to lay telegraph cables across the Atlantic Ocean. His ideas may also have cheered those looking for moral metaphors in science, who perhaps found the equation between work and heat instructive to the laboring masses—at least until the concept of entropy, embodied in the second and third laws of thermodynamics, came along and showed that all work is ultimately futile.

Some years later, with little headway having been made, Thomson devised an analogue computer that he called the Tidal Harmonic Analyser. The device, adapted from technology developed for telegraphy, was an arrangement of a disc, cylinder, and sphere of brass placed in mutual contact—to modern eyes, it looks more like an abstract sculpture than a calculator—that was able to separate out the first two harmonic constituents from raw data by mathematical integration effected by the mechanical rotations of the brass parts. "The object of this machine," Thomson wrote, was "to substitute brass for brain in the great mechanical labor of calculating the elementary constituents of the whole tidal rise and fall." By ganging a number of these contraptions together in a row, Thomson was able to build a machine—fully 20 feet long—that could perform calculations of five major tidal constituents.

Later still, George Darwin took the project over from Thomson. The second son of Charles Darwin, George was interested primarily in the geological history of the earth. He judged a fundamental knowledge of the tides essential in order to fill in the scientific picture of the early earth when both solar days and lunar months were believed to have been much shorter, and the tides far greater. By 1883, Darwin was able to provide an analysis of all the tidal harmonic constituents required for the preparation of tide tables that would be sufficiently accurate for any practical purpose. (A greatly simplified list appears in the table on pages 240–41.) The overall number of contributions is large—it took most of the Latin alphabet and a good part of the Greek one as well to find symbols for them all. The D-Day landings were planned on the basis of tidal calculations that took into account eleven harmonic constituents; modern tide tables factor in anything from sixty to a hundred constituents.

Once the job of analyzing the tides was mechanized, it was only a short step to consider constructing machines that could predict the tide. William Thomson designed one of the first of these contraptions in 1872, capable of taking into account ten major tidal harmonic constituents. Eventually, Thomson set up his own company to build these complex machines, which were exported to many national governments overseas.

The charm of the analogue machines of the Victorian age is that they often reveal something intrinsic about the physical quantities they handle. Unlike the inscrutable digital electronic devices we use today, they show as well as tell. A weighing scales plainly deals with the weight of things. Moving clockwork makes visible the regularity of the passing of otherwise abstract time. Tide prediction machines likewise have the power to give visual expression to what might otherwise seem the dry mathematics of harmonic constituents.

I had hoped to see one of these original tide prediction machines when I visited the National Oceanography Centre's laboratories in Liverpool, where many of these devices were built. But I was told that the prize machine was currently lying in pieces in a back room at the city museum, where it was being restored. Instead, I finally saw one of its sisters in the entrance lobby of the Norwegian Mapping Authority Hydrographic Service in Stavanger. It was a Liverpool machine, dated 1939, ordered before the Second World War but not delivered until after the war was over. The mechanical design is essentially unchanged from Thomson's first machines.

It is a beautiful device, housed in a long, glass-fronted mahogany cabinet that takes up a whole wall of the room. Inside are rows of glittering brass gear wheels facing edge-on to me and spaced along a pair of crankshafts. Above them, a corresponding series of pulley wheels is connected to the gears below by fine metal rods. These are side-on to me and look like larger versions of the wheels inside a wristwatch. Threaded around each of the pulley wheels in turn is a metal wire. It is fixed at one end, while the free end is attached to a weighted pen, which touches a roll of paper placed around a recording drum. The whole mechanism seems to express the very idea of precision. More to the point, I can at last see vividly how the actual rise and fall of the tide in real places is built up from a number of perfectly cyclical oscillations, as the experts have assured me they are. The gears compute the relative contribution to the tide of the various celestial bodies—actual and *fictifs*—that exert a distinct gravitational force, although they also seem to me, in their shining circularity, to symbolize their hovering presence in the sky. The rods connecting the gears

MAJOR HARMONIC CONSTITUENTS OF THE TIDES

CONSTITUENT	SYMBOL	PERIOD	MANIFESTATION
Lunar semidiurnal	M_2	12.4 h	This constituent is what most of us think of as the tide, which is high (or low) about twice a day. It typically represents about half the tidal range we see, with the other half being accounted for by the constituents below, present in varying proportions. Why, if it is the main tide, is it not called M_1? Because the subscript number in this and the other symbols below denotes the approximate number of tidal cycles in a day.
Lunar overtides	M_4, M_6, etc.	6.2 h, 4.1 h, etc.	Analogous to higher-pitched musical overtones, these constituents describe the harmonic "overtides" of the main lunar tide. They contribute to the frequently observed asymmetry of tides in shallow waters, seen in extreme form in river bores, where the tide rises with great rapidity but falls more gradually. The tidal "stand"—a prolongation of the period of high water seen in places such as the Solent and Southampton Water—is due not, as was once thought, to the fact that this body of water has two entrances filled by the tide at slightly different times, but to the near coincidence of the main tide with these overtides.
Lunar elliptic semidiurnal	N_2	12.7 h	This constituent accounts for the fact that the moon's orbit is elliptical. It helps explain why some spring tides are higher than others, because tides when the sun and moon are in alignment and the moon relatively distant (apogee) will be lower than those when the moon is at its closest approach to the earth (perigee).

CONSTITUENT	SYMBOL	PERIOD	MANIFESTATION
Lunar diurnal	K_1, O_1	23.9 h, 25.8 h	These constituents account for the fact that the moon's orbit lies in a plane tilted away from the plane of the earth's rotation, and that the earth's rotational plane is also tilted away from the plane of the earth's orbit around the sun. We see this in the changing height (declination) of the moon and sun in the sky during the course of a day. These constituents make the two high (and low) tides each day usually somewhat unequal in height, especially in temperate latitudes.
Solar semidiurnal	S_2	12.0 h	Because the sun's tidal pull on the earth is just under half (0.46) that of the moon's, this solar constituent generally makes a relatively small contribution to the overall tide. This cycle moves in and out of phase with the major M_2 lunar semidiurnal tides over a period of about fourteen days, which we experience as the constant change between spring and neap tides. The year's highest spring tides often occur around the equinoxes (March 21 and September 23), when S_2 is at its greatest. In parts of the ocean where the lunar constituent happens to be small, this constituent becomes the major cause of the tide, which is then high around the same time every day.

to the pulley wheels convert these circular motions into push and pull motions that provide a kind of diagram of these individual forces.

Each gear wheel along the crankshaft is a different size, and I see they are labeled with the conventional symbols for the tidal harmonic constituents: M_2, S_2, N_2, K_2, V_2, μ_2 and so on. They are geared to rotate at a rate proportional to the period of each harmonic constituent. (The most elaborate analogue tide prediction machine ever built, for the German Hydrographic Institute in Hamburg, took into account sixty-two harmonic constituents of the tide.) As I turn the handle, the gear wheels rotate at their ordained rates, driving the pulley wheels above up and down in a purely harmonic motion. Each movement of a pulley wheel contributes a small alteration in the overall displacement one way or the other of the wire carrying the recording pen. In this way, and without any explicit "calculation," the pen automatically traces on the paper roll the correct sum of all these influences for each instant in time.

Using such machines, predicting future tides for a given port became a matter simply of setting the spacings of the gears to reflect the relative contribution of each harmonic constituent to the overall tide in that place, known from years of historic tidal measurements. The pen then drew out the graph of the tide height on the paper roll for as far into the future as the operator chose to turn the handle. Unfortunately, after that it was down to manual labor to transcribe the graph readings into the more convenient form of tables of tide heights and times. For a different port, it was simply a matter of altering the settings and cranking the machine again. "It was boring work," Tor Tørresen, a senior engineer at the Stavanger office, tells me with feeling. Looking at his white beard, I presume he well remembers the days—as recent as the 1970s—before modern computers took over this task.

SECRET LANDINGS

PORT-EN-BESSIN, NORMANDY

	LW	HW	LW	HW
June 5, 1944	03:45	09:00	16:06	21:19
	1.8 m	6.5 m	1.7 m	6.8 m
June 6, 1944	04:32	09:35	16:52	21:52
	1.6 m	6.7 m	1.6 m	7.0 m

The enduring utility of these splendid Victorian machines was demonstrated to great effect during the D-Day Allied invasion of Normandy in June 1944. The Normandy beaches were well defended with tank traps and other obstructions placed in the hope that they would force any seaborne invaders to approach at low tide, obliging troops to advance for a long way up the gently shelving beach, where they could easily be picked off by German snipers. However, the range of the tide along the northern coast of France is considerable, and the Allies calculated that they could use an exceptionally high and rising tide to carry thousands of landing craft over these obstacles, enabling troops to disembark into shallow water already high up the beaches. The landing craft would then withdraw to safety while the tide was still up. Such a plan relied critically on knowing the precise time of high tide at each of the beaches that would be used, since arriving too early would mean a wait offshore, which might lose the element of surprise, and arriving too late would mean the tide would already be ebbing, making it harder to advance up the beach and risking snagging on the German obstacles. Furthermore, if the landings had to be postponed for any reason, the next opportunity would not come for two weeks, with the next high spring tides, by which time the necessary element of secrecy might well have been lost.

Surprisingly, an 1872 tide prediction machine built to Kelvin's own design was used for this vital war work. The publication of tide tables had been banned in Britain during the war so as not to provide assistance to any German invasion plans, but the data for all the usual ports continued to be processed to produce secret predictions for years ahead. Arthur Doodson at the Liverpool Tidal Institute (a forerunner of the National Oceanography Centre) was the world's leading authority on tide prediction, the heir to Kelvin and Darwin in the previous century. In early 1944, when an Allied invasion of France began to look possible, Doodson was given harmonic data for various new locations known only by code letters, in order to produce tide predictions as demanded by the invasion planners.

As is well known, General Dwight Eisenhower ordered that the invasion be delayed by a day to give time for the seas kicked up by a recent storm to abate. Less well known is the fact that the analysis of the tidal conditions had provided a window of opportunity for the invasion of just three days. The fifth, sixth, and seventh of June were the only days when the tides were judged suitable, being high enough and also rising during the predawn hours, when it was hoped the landing craft could approach the beaches undetected by the Germans. Diary notes made by Field Marshal Erwin Rommel, who had been placed in charge of Germany's "Atlantic Wall" defenses, indicate that the delay may have helped seal ultimate victory for the Allies, because the German commander believed that high tide after the fifth would fall too late in the morning for an invasion to be attempted; so confident was he that the sixth would be impossible that he absented himself from the area on that day in order to celebrate his wife's birthday.

On the Allied side, however, not everybody fully understood the key role of the tide in the plans, in which "H-Hour" of D-Day was in fact staggered to allow for the slightly different times of high water on each of the beaches. According to Antony Beevor's authoritative history of the day, General Leonard Gerow, the commander of the American-army V Corps, which would land on the beach code-named Omaha between Isigny and Bayeux, had wanted to begin the operation at low tide, in order to give his men more time to clear paths up the beach through the German obstacles under cover of darkness. Gerow was supported in this preference by his men, but Eisenhower and the other Allied leaders insisted on attack at 06:30, well after dawn, when the tide would be flooding strongly.

The first of the Omaha landing craft set off at 05:20, facing a journey of an hour or more in seas still running a heavy swell from the earlier storm. Many of the soldiers were seasick, and their vessels soon reeked of vomit. A dozen craft were swamped or capsized, but the rest disgorged their men and tanks 5,000 yards from the shore, stopping

far enough out that they would not run aground. Twenty-seven out of
thirty-two supposedly amphibious Sherman tanks were immediately
lost. Many tank crews and infantry soldiers who couldn't swim were
drowned in the waves. "There were dead men floating in the water
and there were live men acting dead, letting the tide take them in,"
recalled one soldier. One landing-craft commander, Sublieutenant
Hilaire Benbow, did not receive the 06:30 signal and arrived late at
the obstacles, when the tide had risen further. "We were supposed to
have landed when the sea was seaward of the obstacles, all these poles
and crosses and so on, but with the delay the tide had gone in amongst
them so to land the troops we had to get in amongst the obstacles."
Benbow's vessel ran aground on a sandbar, and his men sought refuge
in the water, neck deep, where they briefly felt safe.

Meanwhile, demolition units began to clear the obstacles, working
against the clock under German fire. "As the tide rose, we raced from
one to another," one soldier reported. "One minute the surf was round
their ankles, the next minute it was under their armpits," according to
another eyewitness. They managed to clear a 30-yard gap for the later
landing craft, before the tide forced them to stop; many more gaps
had been the plan, but now the water was covering the obstacles. The
Germans defended their positions tenaciously, and more than two
thousand Americans were killed during the advance up the beach.

Omaha was the second of the five beaches chosen for the Allied
assault. Further to the west was Utah, the other beach where American forces would land. East of Port-en-Bessin and Arromanches were
the British and Canadian beaches—Gold, Juno, and finally Sword—
where high tide came more than an hour later than it did at Utah.
Operations went much better here. At Utah Beach, the rising tide
played a fortuitous role, sending the landing craft more than a mile
south of their planned landing positions, to a part of the beach that
was less heavily defended. No vessels were lost in the calmer conditions here. At the eastward beaches, the sea state was similar to that

at Omaha Beach, but the order to launch the tanks at thousands of yards out was ignored; instead, the landing-craft captains were able to come in closer to the shore and lost far less matériel.

"The D-Day landings might be considered the ultimate success for those big, beautiful, brass tide-calculating machines," according to the American tide expert Bruce Parker. The tide predictions were also a victory for expert intuition and experience. The Allies lacked the necessary harmonic constants to set the Liverpool tide prediction machine to produce accurate tide times for the actual Normandy beach sites. They had data only for the major ports of Cherbourg and Le Havre, some 30 miles away in each direction along the coast. (Port-en-Bessin, the location for which I gave the relevant tide times at the start of this section, was at that time a secondary station that simply estimated its tide times by subtracting a certain number of minutes from the times for Le Havre to the east, and took no account of any local variations.) Earlier undercover reconnaissance of the beaches in small boats and submarines—exercises that surely must have evoked thrilling memories of *The Riddle of the Sands* for many of those involved—proved unable to furnish the extended series of tidal measurements that were needed to make future predictions. In the months before D-Day, the Admiralty sent Arthur Doodson repeated requests to plug the mysteriously placeless harmonic constants into his machine so as to provide reliable tide tables; he was never told the location of the planned landings. After the war, though, Doodson admitted that, despite the secrecy, he had been able to guess the site solely from the harmonic constants supplied to him.

It is not only in places where the tides are substantial, as they are on the Normandy coast, that they need to be taken into account in planning amphibious military operations. When the Americans went to capture the atoll of Tarawa (one of the Gilbert Islands, now called Kiribati, in the equatorial Pacific Ocean) from the Japanese, the operation was almost a disaster. In the Pacific theater, the US Coast and Geodetic Survey was the body charged with supplying tide predictions, but the distances between locations for which there existed reliable data were far greater than on the Atlantic seaboard. On Tarawa, as on many Pacific islands, the tidal range is never more than a couple of meters or so, and one might be forgiven for thinking the tides could be discounted in planning a landing. The Americans nevertheless made due efforts to collect relevant tidal information in advance of the assault in November 1943. Unfortunately, the nearest port for which they had good data was Samoa, 2,000 kilometers distant, far enough to render any comparison meaningless. Other tidal stations were just as far-flung, and often in places where the tides were anomalous anyway.

After various delays, the assault was finally set for the twentieth, which happened to be the day after an apogean neap tide—one of the days of the year with the smallest range of tidal movement, when the moon is at its most distant from the earth and when the gravitational attractions of the moon and the sun are acting against each other.

The strategy involved taking the island from the side where it was less heavily defended, where there was a lagoon within a coral reef. The amphibious assault would have to cross over the reef in order to enter the lagoon. As at Normandy a few months later, an early-morning landing was thought advantageous. This window of opportunity was predicted to have the rising tide needed to carry the troops over the reef, and it left time for a second wave of troops to advance on the following high tide that evening while it was still light. Reliable local informants warned the Americans that the tides that day would range hardly more than a meter, meaning that they would risk grounding their landing craft on the reef. Holding off for a few days would bring the return of higher tides—an option favored by some of the men on the spot—but higher up, the decision was made to press on. The marines reinforced the bottoms of their vessels with metal armor and waited for the order to go in. That morning, an advance fleet of shallow-draft vessels successfully carried the first troops over the reef, but the following transports, with deeper keels, stuck fast on the coral, forcing the marines to disembark and swim or wade through the lagoon under heavy Japanese fire. The vessels did not finally float free until three days later, by which time they had been shot to pieces. The Americans lost more than a thousand men taking the tiny island.

In disbelief that the tides could have let them down so badly, the US Navy spent the months on Tarawa after the invasion making new tidal measurements. Their readings confirmed that the predictions they had used were substantially accurate. With such a low high tide, however, it was possible that the precise time of high water, and the path taken by the flow and ebb either side of it, might be substantially affected by nontidal factors, such as local and temporary weather effects, creating what locals called a "dodging tide." Some retrospective accounts of the Battle of Tarawa suggest that there was afterward an inquiry into the "failure" of the tide that day; others insist that no such investigation was ever ordered. A biography of Rear Admiral Richard Turner, who was in charge of the assault, makes a desperate effort to exonerate its subject: "Like other sudden variations in natural phenomena, 'man proposes, God disposes.' The tide suddenly and dramatically failed."

10

NATURE'S
FREE RIDE

KNOTS LANDING

In winter, the knots come by the thousands to feast on the riches in the intertidal mudflats of the Wash. These medium-size wading birds are rather dull and unprepossessing in appearance, white underneath and sandy on their backs. During the breeding season their plumage turns a rich russet, but this happens in the Canadian Arctic and other colder climes where the birds spend the summer, and we seldom see it.

In Britain, knots are more remarkable for their sheer numbers. Vast flocks of them put on a mesmerizing aerial performance over their feeding grounds, here in my county of Norfolk, in the Severn and Thames estuaries, in Morecambe Bay, and in other areas where sufficiently broad stretches of mud are exposed by the tide. Witnessing this behavior requires having a tide table on hand, for it takes place only toward the end of the flood tide as the waters rise to cover the last of the mud. The birds then pass the hour or two of high tide hidden in vegetation on slightly higher ground before returning to feed once more. At Snettisham in Norfolk, on the west-facing shore of the Wash, high-tide flights of forty-five thousand birds have been recorded.

The knots' behavior is highly distinctive. Some wader species fly off at intervals as individuals when the tide comes and goes. Others, such as oystercatchers, turn their back on the water as the tide advances, and walk disconsolately, it seems, up the mud slope in time with the rising sea. But knots, perhaps because they cluster too tightly to move in this way, face the rising water and take off at the very last chance they have, when it has risen up their legs and threatens to wet their underfeathers.

I watch as they fly off in a continuous flourish like a magician's curtain being swept aside. The dense flock swirls and feints like a single aerial organism. Once aloft, the birds first coalesce as an egg-shaped cloud low over the water, before gaining height and taking on ever more extravagant, twisted shapes like a pixelated flamenco dancer. It is a spectacle, but it is not a display in the biological sense; it is a defensive behavior, generated by each bird's instinct to find a position securely within the flock, away from the edge where it might be picked off by a peregrine falcon. Perhaps the single shifting mass that so captivates us as spectators appears as a single terrifying creature to these predators. As the ever-changing shape swerves and dives, each of its dots switches between light and dark as the bird's body banks into a turn, creating a scintillating effect in the smokelike apparition. All the while, the birds call, *queek-eek*, their cries piling up on one another to make a deafening, high-pitched white noise.

The knot fully lives up to the species name given to it in 1758 by the Swedish naturalist Carl Linnaeus: *canutus*. It entirely fails to halt the incoming tide. Once it has rested ashore during the hour or so of the highest tide, it returns to its task, continually marching down to the tide's edge as the water recedes, in search of food. The returning flock is less theatrical than the one that rose up into the sky a short while ago, and more urgent. Now a constant rush of birds streams out low over the marshes like accelerated particles to settle on the freshest mud at the instant it is uncovered. The knots spread themselves out along the ever-changing edge of the water. These birds are tactile feeders that use their long bills to detect vibrations transmitted through the mud that reveal the presence of buried mollusks. They eat small mussels and cockles but are especially partial to a species of clam known as the Baltic tellin, *Macoma balthica*, and a small mud snail, *Hydrobia ulvae*. The tellin lives in the mud of the lower part of the intertidal zone and down below the mean low-water line, where it uses a siphon to draw nutrient material down to it from the wet surface. The mud snail, on the other hand, inhabits the upper reaches of the intertidal zone and feeds on seaweeds—it is also known as the laver spire shell—and organic detritus, such as the decaying remains of sea creatures and fecal matter, from which it extracts protein. This segregation of its principal foods is reflected in the knot's feeding habits—pecking from the surface for snails in the higher parts of the intertidal zone when the tide is approaching high water or beginning to ebb, but probing and pushing in the mud for tellins in the lower reaches when the tide is out.

Other shorebirds favor different foods and therefore exhibit somewhat different behaviors in relation to the water. Where the knot stays Canute-like at the edge, the dunlin is happy to wade into the water after its prey. Avocets use their upturned bills to skim very shallow water for rag worms and shrimp. The oystercatcher can break into the harder shells of limpets using its jackhammer of a bill. I have even heard that unwary birds are occasionally caught in the viselike grip of the shells of their would-be prey and mercilessly held until the tide returns and they are drowned. In the Bay of Fundy, up to two million semipalmated sandpipers stop off on their migration to feast on mud shrimp, the location favored because the invertebrates develop earlier there owing to the exceptional tidal range, which in turn guarantees a more reliable food supply.

In all cases, the birds' behavior is regulated primarily by the tide. Indeed, animals such as these birds and their various prey species are often critically dependent on these habitats, which makes proposals to build barrages and other interventions that might affect the tides especially worrisome. The ritual of the knot—the feeding, the flocking, the roost, the return to the feeding grounds—is repeated, not at the same time, but nearly an hour later each day because it is driven by the cycle of the tides. When, perhaps after several days, the high tide comes in as night is falling, the knots can afford to roost for a little longer and can dispense with the aerial acrobatics, because their predators hunt by day, but they carry on feeding during the nighttime low tide as well as the daytime low tide. Species with longer legs and longer bills are able to feed in slightly deeper water and are therefore less constrained by the tidal cycle and more likely to exhibit a daylight feeding pattern. Exclusive daytime feeders will come to the shore if the tide is right but may otherwise seek a varied diet on fields inland.

The intertidal zone is a unique habitat, a concentrated band of transition between the ecological communities of the sea and the land. It is an environment of high rewards in terms of food—Strabo long ago observed how the mussels on the Atlantic coast of Spain grow larger than their Mediterranean cousins—but also of high stress, owing to exposure to heat and cold, sun and waves. Other habitats tend to be confined to climatic regions. But the intertidal zone is present on every coast, delimited only by the tides. The kind of life it supports is closely replicated all around the world too, notwithstanding differences in climate.

In the muddy inlets of Norfolk, I found subtle changes in vegetation and minerals as I moved up and down the shoreline. But this banding is even more pronounced on rocky shores where the division into zones is precisely painted onto every surface, offering pictorial evidence that animal and plant species are rigidly organized into tiers according to their tolerance to immersion in seawater or exposure to the air. On such a shore I find the topmost stratum is essentially terrestrial rock, a clean midgray in color, with green moss growing on it, and splotches of guano showing that the rocks are used as regular perches by seabirds. Below this layer is a darker-gray stratum, more precisely delineated at the top than at the bottom. This is the splash zone, exposed to frequent dousing in brine but not to regular immersion; the different color indicates the presence of small algae.

To the lower slopes of this stratum, in places where it is not steep, clings a wrack-type seaweed. Below this is a brown layer comprising more algae and seaweed, which is covered at every high tide. The final layer I can see, toward the low-water mark, shows gray rock once more, with larger marine organisms such as barnacles and limpets and larger seaweeds fastened to it. It is not only these fixed, or sessile, species that exhibit zonation, but also the myriad insects, crabs, shorebirds, and other creatures that scurry up and down in search of food within their own favored limits.

The vertical sequence is the same the world over, although the width of the individual bands may vary, and sometimes one layer may be squeezed out altogether because the physical conditions are not right or because of competition between species. In general, however, this hierarchical organization of life acts to reduce such competition. The overall extent of the zone is determined by the amplitude of the tides, with the upper limit of colonization determined by tolerance to dry conditions and the lower limit defined by the incursion of predators such as starfish that feed on mollusks only up to the low-water mark because they themselves are unable to survive out of water. The bands are as rigid as any national border, and may be shifted only by an unfortunate coincidence of events, and then usually only a little, and for a short time. Battering storm waves on a high tide, for example, will temporarily increase the extent of the splash zone, while a spell of drought occurring at neap tides may dehydrate species that normally live near the mean high-water mark.

There is inevitably one other major threat to the intertidal zone, and that is humankind. In many parts of the world, this rich habitat is being severely squeezed. Rising sea levels encroach from the sea side—a more significant factor than it seems: a single millimeter rise in sea level can be responsible for erosion reaching up to a meter inland. Meanwhile, coastal property development eats away at it from the land. If the two meet, sea defenses may be constructed, replacing wide mudflats, a beach, dunes, and marshes with one hard, concrete barrier. A band of habitat once a mile or two in extent is suddenly supplanted by a more or less vertical wall that compresses the intertidal zone into a few yards.

THE GRUNION RUN

LA JOLLA, CALIFORNIA

	LW	HW	LW	HW
April 22, 1947	03:45 −0.3 m	09:58 1.2 m	15:12 0.3 m	21:31 1.8 m

Other creatures have evolved to use the tides in more sophisticated ways. The breeding cycle of the California grunion, for example, is synchronized exclusively to the spring tides, the highest of high tides that arrive at fortnightly intervals on each new and full moon. The grunion is a small, silvery fish about the size of a sardine with ray-like fins. On spring and summer nights, when the tide is high, the fine sandy beaches of southern California where they come to spawn are suddenly covered in thousands of the writhing creatures. First come the males, holding themselves back in the surf until one wave comes that will pitch them high up the beach. From there, they propel themselves further up the beach using their fins and tails in a wild corkscrew action to cross the wet sand. They are soon joined by the females, who excavate shallow nests somewhere in the top few meters of the zone of the beach that is washed by these high-tide waves. The eggs of each female may be fertilized by the milt of several males, which afterward make their way swiftly back to sea. The female then digs her tail into the soft sand and lays her eggs before she, too, returns to the ocean. Mature females are able to spawn in this way several times at fortnightly intervals in timing with the spring tides, and the season may last from March to August.

Often, the number of beaching fish is so great that they must crawl over one another in the race to get up the beach and then safely away before the tide recedes. The "grunion run" has become a popular spectacle, and wildlife enthusiasts sometimes crowd the nighttime beaches almost as densely as the grunions themselves.

The phenomenon has also been much studied, not least because one of the beaches where it happens lies directly in front of the Scripps Institution of Oceanography at La Jolla. "It is truly amazing to see hundreds of fish on the beach, and then five minutes later to be unable to find any," wrote Boyd Walker, a marine biologist at Scripps, who made a pioneering study of the grunion run in the late 1940s.

Scientific observation has revealed that the event is precisely attuned to take advantage of the tides. The run occurs mainly on the days immediately *after* the full or new moon, when the tides are expected to be high, but not quite as high as the very highest spring tide. The high tide on each successive day during this short period will be a little lower, helping to ensure that the new-laid grunion eggs will not be washed away. In fact, the breaking waves of the subsequent tides throw up sand that helps to protect the eggs by burying them gradually deeper. Over the following eleven days or so, the eggs are incubated in the shallow, moist sand. Both the sun's warmth and the moisture are important for incubation: a Gulf of Mexico species of grunion is adapted to spawn in the mid part of the intertidal zone because the higher reaches of sand on its beaches are prone to drying out. After eleven days, the first big waves of the next spring tide arrive. The waves begin to erode the beach where the eggs lie buried. Soon the eggs are no longer snugly encased in sand but are being shaken about in the waves, which is their signal to hatch into larvae. Typically, the eggs hatch very shortly *before* the next new or full moon, giving the larvae the greatest chance of making it down the beach and out to sea over the next few spring tides, which will be of increasing height.

This is not all. If the following spring tides happen not to reach the incubating eggs—for example, because offshore winds reduce the height of the tide—the fertilized eggs are able to delay hatching for two weeks or even four weeks until a more propitious tide comes. The variable period of incubation is a useful survival strategy in a physical environment subject to unpredictable irregularities. The grunion is also adapted to make best use of the pattern of the tides specific to its California habitat. The tides along this coast exhibit a strong semi-diurnal inequality, which means that the two tides each day are often rather different in height. During the winter, the highest tides are usually during the daytime, but in summer they come at night. This is why both the spawning and the hatching take place under cover of darkness, affording the fish and their larvae a measure of protection from predators as they struggle up and down the beach.

A number of marine species, including whitebait, mummichogs, and Colchester oysters, exhibit a similar, though less marked, synchrony with the spring tides in their breeding cycle. Fiddler crabs work their way down the shore, foraging in the spume when the tide is ebbing, before sealing themselves in their burrows when the flood comes. Their reproduction is governed by the lunar cycle, with mating and spawning occurring on successive spring tides so that the new-laid eggs are dispersed by the sea. But in the grunion, the adaptation is so fine-tuned that it is considered an evolutionary marvel. "The adjustments are many and precise, but only to the special tide and beach conditions that are encountered," observed Boyd Walker. The evidence strongly suggests not only that the grunion evolved in this spot, but also that the tidal conditions of its habitat have remained substantially constant ever since.

From the intricacy of its mating cycle, it is clear that the grunion is engaged in a subtle dance with the tide. But some other creatures that might seem at first to exhibit behavior related to the tides may be responding to the changing light of the moon. For example, brain corals grow new skeletal layers more rapidly when the moon is full. The thickness of the layers linked to the lunar cycle can be used to establish their age, like tree rings. Some old corals show annual and daily growth bands, as well as those related to the lunar month, from which paleontologists have been able to deduce that, in the Devonian period four hundred million years ago, there were four hundred days in a year, each lasting about twenty-one hours, because of the earth's faster rotation at that time.

There may even be natural responses to far longer cycles, such as the so-called lunar standstill, which happens at regular intervals of 18.6 years, when the angle of the earth's axis and the angle of the plane of the moon's orbit combine in such a way as to maximize the height of the moon in the sky. For example, mussels cluster together to form intertidal beds that long outlast the lifetime of individual organisms, which itself may be twenty years or more. The upper limit of these beds has been observed to move up and down in time to this long cycle in response to the amount of time the mollusks are able to spend out of water.

In the Arctic, it is easy to appreciate that survival may depend on a sensitivity to fluctuations that would be regarded as insignificant anywhere else, and one might therefore expect to observe such subtle long-term connections in this slowed-down environment. In the 1930s, a Danish zoologist called Christian Vibe—a name now more suggestive of an evangelical radio station—traveled to Greenland determined to know more about the mysterious long cycles that seemed to govern the life there of the caribou and the cod, the hare and the herring, and to discover what they might mean for the earth's climate as a whole. He examined the population cycles of various Arctic creatures along with fur traders' records and other data, and believed that he had found a link between their migrations in latitude and the 18.6-year cycle, although the claimed correlation remains contentious. With increased interest in climate change, there are more intensive studies of animal population variations today, but in the end there are so many overlapping cycles to consider—the eleven-year period of the sunspot cycle is another—that it can be hard to be persuaded of the overriding importance of any one factor.

THE RHYTHM METHOD

Numerous other animals observe the cycle of the tides. Long before she wrote *Silent Spring*, the book that would bring her lasting fame for sounding an early warning about damaging human impacts on the environment, Rachel Carson was a marine biologist. Her earlier books were all about this environment, including one—*Food from the Sea*—devoted entirely to its bounty that might raise a few eyebrows today, when there is strong evidence for believing that stocks of many fish are endangered. Her first book was born when the inherent poetry of her writing demonstrated its unsuitability for the dreary fishery brochures she had been employed to write. *Under the Sea-Wind* was published in 1941. It displays a degree of anthropomorphism perhaps unacceptable in science writing today, as Carson follows the progress of individuals of various species as they make their lives in the sea. She even gives them names, as if they are characters in a timeless fable.

Carson made many of her observations on the barrier islands of the Outer Banks of North Carolina, where there is now the Rachel Carson Reserve. She describes the black skimmers ("Rynchops") known locally as "flood gulls" for their habit of feeding on the rising tide that brings small fish into the shallows, which they scoop from the water with their massive red bills held agape as they fly along. Like the knots, these birds feed on every tide, even when it means hunting in darkness. The killifish, which are the skimmers' chief prey, in turn have used the tide to catch a free ride into the marsh-grass roots where they are safer from the birds, which cannot fly low over the grass. They feed on insect larvae that, conveniently for them, have also been lifted into the water by the rising tide. From terrapins to sea fleas, Carson paints an intimate group portrait of a subtly interlocked ecology, all driven by the rhythm of the ocean tide.

Humans, too, play a part in this drama, or have done so, for methods that respect the natural cycle are increasingly being swept aside by the blunt instruments of industrial aquaculture. Like many other creatures, we enjoy the easy harvest the tide brings. Cedric Robinson, the Queen's Guide to the Sands, spent his boyhood shrimping and cockling in Morecambe Bay. In the softer mud of Bridgwater Bay in Somerset, Adrian Sellick is said to be the last man to fish using a mud horse, a sled-like device that allows him to slide out across the shallows where the shrimp are caught. The technique is centuries old; now he blames the looming nuclear power station at Hinkley Point for altering the water flow, ensuring that he will be the last of his kind. The remains of structures found under the sea indicate the historical importance of the intertidal zone for trapping fish too. These traps have been found made of wood or rocks around many coasts. The northwest coast of North America is especially rich in them, but the oldest, found in the Baltic Sea, is believed to date from about nine thousand years ago, placing it roughly concurrent with the dawn of agriculture.

Seaweed is frequently taken from the intertidal zone. Farmers in Tanzania, for example, have turned to the seashore to grow red *Eucheuma* seaweed for carrageenan, which is used as a gelling agent by the global food industry. Even in my corner of East Anglia, there is serious talk of growing seaweed for biofuel, as well as for high-value uses as food and as a source of chemicals that might be developed into pharmaceuticals.

It is not only fishermen and farmers who used to work tide-governed shifts, as I realize when I visit the tide mill at Woodbridge in Suffolk. There are records of such a building going back to 1170, although the version I see was rebuilt in 1793; it only ceased commercial operation in 1957. As a working model shows, the usual mill machinery is operated by using the head of water created when the tide is held back in a pond adjacent to the mill. The controlled release of the water from this pond gives the miller four or five hours during which to grind flour. Of course, this opportunity comes twice in every twenty-four hours, causing the miller to refashion his daily schedule as much as any knot or skimmer.

At the mill, I see a reproduction of a medieval illustration testifying to the ancientness of this technology. The image, from a Lincolnshire document called the Luttrell Psalter, dating from around 1330, has the precision of a scientific diagram and shows clearly the head of water produced in the pond and the channel that leads it to drive the waterwheel on the side of the millhouse. Under the salt water in a corner of the pond, the artist has included a couple of elaborate wicker baskets of a type designed to trap eels, whose life cycle carries them from the remote ocean where they are born to the rivers and eventually back to sea to breed, using, as Carson writes, "the large and strange rhythms of a great water which each had known in the beginning of life."

TIDE SENSE

How do animals "know" what the tide will do? When scientists began to study them, they thought it unlikely that grunions, for example, could have a wholly internal mechanism by which to time their breeding run. Do they then sense the tide directly, or are they responding to other stimuli? Certain things can be deduced. It cannot be moonlight that causes the fish to run ashore, as they do this at high-water springs both on the bright full moon and in the darkness of the new moon. The fact that spawning occurs not on the very highest tides, but shortly afterward, indicates that the behavior is not directly governed by tide height either (something that the grunion might detect as increased water pressure around its body). It has recently been found that it is agitation in the sea when the *next* spring tide comes that stimulates the grunion eggs to release their larvae by generating enzymes that dissolve the egg membrane. But wave action cannot be the trigger for the original beaching, or else storms would disrupt the pattern of the spawning runs.

What else is left? Could the stimulus be pure gravitational attraction? Any forces felt by the fish would be minute in comparison with the changes in pressure that they tolerate when swimming at different depths, but gravity cannot be entirely discounted. Whatever triggers the behavior, it seems it is not confined to the single tide when the grunions come ashore to reproduce, since the eggs begin to mature long before they are spawned—and this, too, has been found to happen in concert with moon phases.

The author John Steinbeck writes about this conundrum in *The Log from the Sea of Cortez*, the record of a specimen-gathering expedition he made to the Gulf of California with his friend Ed Ricketts in 1940. (The *Sea of Cortez* was the name of his vessel as well as his happy hunting ground.) Ricketts—who provided Steinbeck with the model for the character of Doc in *Cannery Row*—later wrote a scientific guide to intertidal life based in part on what the two men collected on the trip. Tangling with mangrove roots in their motor launch at high water, and racing against time to gather specimens from the uncovered beds at low tide (like knots), they find an astounding diversity of colorful life in the warm waters: crabs and snails, and creatures with names out of horrible myth, such as the gorgonian, and others, like the serpulids, so obscure that their only name is the exotic-sounding one given to them by science. They see fish that can survive out of water, at least for the period of one tide, and minutely observe how the foreshore is graded with different kinds of life by height and by the time spent immersed in seawater.

Inevitably, they find themselves drawn into speculation on the importance of the tide to this abundance of life, especially since, in Precambrian times, when single-celled organisms began to evolve into more complex forms in the sea, the tides were far greater than they are today, because of the closer orbit of the moon. Steinbeck writes, "The moon-pull must have been the most important single environmental factor of littoral animals." Their body weight and displacement in the sea would have cycled strongly with the rotation of the earth. "Consider, then, the effect of a decrease in pressure on gonads turgid with eggs or sperm, already almost bursting and awaiting the slight extra pull to discharge." What Steinbeck finds more remarkable is that so many creatures seem to have carried forward a kind of ancestral memory of this response and fine-tuned it to the far weaker signal of tides now—an effect to which, he believes, even we are not immune. "Tidal effects are mysterious and dark in the soul, and it may well be noted that even today the effect of the tides is more valid and strong and widespread than is generally supposed."

But we are still left with the question of how it is that these creatures respond to the tides. They do not go around with tide tables; they do not superfluously relate tide to time as we do. So either they must have some kind of built-in tidal clock, or they must directly sense some primary or secondary property of the tide, which might include the pressure or rate of flow of water, or its change in temperature or salinity.

Perhaps our own removal from the tides is making the problem seem harder than it is. It is, after all, no miracle that animals are sensitive to time. We ourselves are slaves to the circadian rhythm, as our alarm clocks and news bulletins daily remind us. Why should a circatidal rhythm be any odder than a circadian one? Of course, the circadian rhythm has the obvious cues of bright sunlight alternating with utter darkness. But this is what is obvious to us. What cues might be just as obvious to marine creatures very different from ourselves? We find a twenty-four-hour cycle perfectly natural, but might not these animals find a period of twelve-and-a-bit hours just as natural, especially if their survival depends on it?

. . .

MORE RECENT RESEARCH has revealed even stranger ways in which marine creatures harness the tide. A Pacific species of starfish, for example, the purple sea star, *Pisaster ochraceus*, has developed a water-cooling system to help it survive accidents of tidal stranding. This curious response is triggered when the animal first becomes hot as the sea leaves it behind on one low tide. This unhappy event triggers a mechanism that enables the sea star to retain some seawater when the tide next rises and then recirculate it when the tide falls once more so that it does not experience the same discomfort again.

Where survival is not at stake, the drive is often to conserve energy—the explanation of so many natural adaptations. There is an equivalence between food and energy: food provides energy, but it costs energy to find food. If the tide brings in a food supply, then that represents a saving of energy, like having groceries delivered to your door. For any sea creature, the tides are a huge source of free energy there for the taking.

This is why places where the tides are large or fast-running are often high in biodiversity. Some creatures draw an energy advantage from the movement of so much water, while others are drawn in to feed on them, building a rich ecosystem. This richness, in turn, is a large part of what gives tidal locations their strong sense of place. These places are usually on or close to the shore, but they may also occur over undersea ridges or shelves where the tide runs in such a way as to mix waters from different depths, churning nutrients and dissolving oxygen from the air. It is striking that the larvae of the eel, whose life cycle had greatly baffled naturalists since Aristotle because they could not determine the sex of the creatures and had never seen their young, were originally detected in the Strait of Messina as late as 1856, although they were at first misidentified as a new species of fish. It seemed that the mysterious larvae were drawn up from the depths by the turbulent Charybdis whirlpool, and it was presumed that the eels spawned nearby in unknown depths. It was another fifty years before the Danish biologist Johannes Schmidt established the vital importance of the Sargasso Sea in the life cycle of the eel, and showed that the Mediterranean was not a breeding ground. The young fish were perhaps drawn to the Messina Strait for its rich food supply.

The flattened variety of sea urchins known as sand dollars may be adapted to take advantage of the tide in another way. Their shells, or tests, exhibit the usual fivefold symmetry of this class of echinoderms, but they are also often pierced with elongated holes, called lunules. One species, which rejoices in the name *Mellita quinquiesperforata*, has five of these holes, but others may have different numbers, and in different places on the shell, not governed by the overall fivefold symmetry. The variable shape and number of these features poses a puzzle to zoologists. The lunules do not assist with feeding or offer any structural advantage to the test. In 1981, however, Malcolm Telford, a marine biologist at the University of Miami, compared the way water flowed around a perforated and a nonperforated species in a water tank, and found that when the water was flowing fast enough, the perforated sand dollar was able to channel the flow through its shell in such a way as to generate lift and stabilize its vertical displacement in the water. The lift occurs in water moving at speeds typical of tides in shallow waters, suggesting that the holes might have evolved as an energy-saving strategy, enabling the animals to ascend or descend in the running tide with no more than a slight tilt of the body.

Other marine species use the tide to hitch a ride even more blatantly. Julian Metcalfe is one of a group of scientists at the UK Centre for Environment, Fisheries and Aquaculture Science (CEFAS) interested in fish migration and its implications for the sustainable management of commercial fish stocks. The government organization is based in Lowestoft, formerly one of the biggest fishing ports on England's North Sea coast. It occupies a suitably forlorn complex of brick buildings that began life as a hotel when vacationers were still satisfied with sea views painted entirely in shades of gray. If one wanted a satirically British counterpart to the sun-soaked modernism of the Pacific-facing Scripps Institution, one could hardly do better.

One focus of Julian's research is the plaice, one of the large edible flatfish traditionally landed here. Like many species of flatfish, plaice are known to lift themselves into the tidal stream when it is running in the direction they wish to go and to drop out to lie on the seabed when the flow is unfavorable. The behavior contrasts with bigger, round fish such as cod and tuna, which are unable to burrow into the seabed to escape the tidal currents, and so usually find it more advantageous just to keep swimming. The plaice are perfectly capable of swimming too—and they do so where the waters are not strongly tidal—but they prefer to save energy where they can.

By tracking fish using imaging sonar from the CEFAS research vessel, and more recently by collecting data from electronic tags fitted to individual fish, Julian and his colleagues have been able to map the shifting populations of the fish through the seasons. It emerges that they are adept users of the tide. When the plaice are spawned, the eggs and tiny larvae of the fish are first carried helplessly by tidal currents into shallow coastal nurseries, where they are able to grow at reduced risk of predation. Fish spawned off the Thames estuary, for example, may be transported as new-hatched eggs toward the Frisian Islands off the Dutch coast. There the tide whisks them through the Texel island gate between two of the islands into the shallow and protected Waddenzee, where they grow up to 20 centimeters long. The young fish are able to stay safely within the nurseries by resting on the seabed when the tide is ebbing, and rising up into the water when it is flooding. They do this using a kind of internal tidal clock that is sensitive—or "entrained," in the jargon of the field—to changes in water pressure produced by the changing height of the tide.

When they have grown to adulthood, the plaice are able to migrate between their winter spawning grounds in the North Sea and their ocean feeding grounds, riding on the tide when it is running in a favorable direction and dropping to the bottom when it is not. They are thought to use a combination of both circatidal and circadian biological clocks to do this. The plaice can tell whether a tide is running to a high degree of accuracy, remaining on the seabed during slack water, but lifting off within minutes of its beginning to run their way. But there remain unanswered questions about this behavior: How do the fish know where to go? How do the fish sense the *direction* of the tide? And how, with no change in tactile or inertial sensation, do they know when the free ride is over and the tide that has been bearing them along comes to a halt, especially if it is dark and there are no visual cues?

For Julian, the plaice is a highly convenient fish to study. The technology and resources were made available for research because it is a commercially important species; plaice is third in the trio of popular fish-and-chip-shop varieties of bottom-living fish caught in the United Kingdom, after cod and haddock. But the animal has since been shown to exhibit a migratory behavior that is fascinating in its own right. Julian is convinced that similar mechanisms exist in many species. "I can't believe lots of animals don't take account of water currents," he says. "We can now modify our environment so much, we forget the old cues, the ones that animals still use."

Solar control of the daily and yearly rhythms of life on land is relatively well understood (and of course inescapably familiar in ordinary human experience). But an explanation of the rhythmic mechanisms of marine life is only just beginning to emerge. Circadian rhythms are regulated by genes that provide chemical feedback. This means that the rhythm is maintained in a "free-running" fashion even in the absence of external stimuli such as changing light levels or temperatures. Similar free-running rhythms are seen in sea creatures, but it has been arguable whether these biological clocks are truly tide related or are versions of the circadian clock tweaked by the processes of natural adaptation to operate at a different rate.

In 2013, however, geneticists at the University of Leicester obtained evidence to suggest that a dedicated circatidal biological clock does exist. Researchers led by Charalambos Kyriacou worked with the speckled sea louse, a familiar denizen of intertidal sandy beaches that bears the deceptively alluring Linnaean name *Eurydice pulchra*. By disrupting the expression of genes responsible for circadian timekeeping, and showing that the animals nevertheless maintained their normal tidal behavior, they established that the tidal rhythms are driven independently by a circatidal clock. The lice have biological clocks of both kinds: a circadian clock that governs such matters as the production of pigments in the body, and the circatidal clock that regulates their swimming activity in response to the twelve-hourly cycle between successive high tides.

Perhaps many creatures have the potential to switch over to circatidal rhythm. On the walled island of North Ronaldsay, the northernmost of the Orkneys, the local sheep were excluded from grazing land in 1830 when the laird decided to use the lush grass to breed cattle. Except in the lambing season, the sheep are confined to the beach side of the coastal wall. Over the years, the animals have adapted to a diet of seaweed and a grazing timetable governed by the tides that uncover it.

As for the longer cycle of spring and neap tides (the word "circalunar" is used to distinguish this period from the circatidal response to *each* tide), there is new evidence to explain animal responses to this too, from similar experiments at the Max F. Perutz Laboratories of the University of Vienna. The Austrian researchers used rag worms for their subjects. This was one of the first species observed to spawn on a cycle attuned to the spring tides and is considered a living fossil, with a physiology, behavior, and habitat unchanged over millions of years. Unlike the grunion, it does not spawn on every spring tide at the right time of year, but only monthly, on the spring tides at the new moon. This behavior suggests that the animal's circalunar clock may be entrained to moonlight, or rather the absence of moonlight, and not to the hydrodynamic factors that might be important for the grunion. Biochemical reactions catalyzed by lunar light may play a role in this programming.

All known biological clocks are linked ultimately either to the sun or to the moon. The discovery of circalunar clocks in animals must surely raise the hopes of those many horticulturists and farmers who believe, scientifically or not, that their seeds are better planted and their crops are better harvested at certain phases of the moon. And what about us? The wife-and-husband team who led the Viennese research, Kristin Tessmar-Raible and Florian Raible, conclude their paper with this bold question: "Is it possibly more than sheer coincidence that the female reproductive cycle in humans lasts around a lunar month, or could this instead reflect some regulatory left-over from our evolutionary past?"

I had, of course, been entirely unaware of the dependence of the rag worm on the lunar cycle when I picked one out of the mud during my day observing the tidal cycle on the north Norfolk coast. I saw gulls and curlews come and go that day too, but I would have had to stay for many more days in order to appreciate all these animals' obeisance to the tides. It is surely a mark of how much our own lives are ruled by the black and white of day and night that we are so insensitive to these different rhythms, and it is only now that science is beginning to comprehend them.

POOLING OUR KNOWLEDGE

With this abundant evidence that life depends in many ways upon the tides, we are entitled to ask whether life could have evolved on earth at all in their absence.

Charles Darwin has little to say about the tides in *On the Origin of Species*. He uses the word only three times—twice in a metaphor, and once in a geological argument in relation to the erosion of cliff bases. Yet by the time the book was published in 1859, and possibly for some time before that, he firmly believed that life must have had an aquatic origin. However, he forbore speculating on the matter in print, publicly saying it was "rubbish" to think of doing so, and it was only much later, in a short letter to the botanist Joseph Hooker in 1871, that he made the remark that soon became famous, imagining the "warm little pond" where the right simple chemicals might first have converged:

> It is often said that all the conditions for the first production of a living organism are now present, which could ever have been present. But if (& oh what a big if) we could conceive some warm little pond, with all sorts of ammonia & phosphoric salts,—light, heat, electricity, &c, present, that a protein compound was chemically formed, ready to undergo still more complex changes, at the present day such matter w^d be instantly devoured, or absorbed, which would not have been the case before living creatures were formed.

The scenario soon captured the public imagination. In *The Mikado*, Gilbert and Sullivan's comic opera of 1885, Pooh-Bah claims descent from "a protoplasmic primordial atomic globule."

Darwin presents in a nutshell the conundrum that still baffles scientists today. For the continuing existence of life may paradoxically be the very thing that prevents us from observing how it emerged in the first place. The fact that life is not observed spontaneously forming on the present-day earth leaves the question of its origin wide open.

The immediate context of Darwin's letter to Hooker was the fact that Louis Pasteur had shown by experiment that it was impossible to generate germs spontaneously, without there first being some other organic matter present. Darwin never expressed a view on how the first germ came about. Among the theories seriously entertained by others were and are that the necessary chemicals were delivered to this planet by comet impacts or, alternatively, that they were thrust up to the surface of the earth from deep within its crust. Several reputable scientists, including Darwin's champion Thomas Huxley, even made excited claims to have found protoplasm in mud from the bottom of the sea.

For all Darwin's stature, a "warm little pond" is not currently regarded as the most likely stage for the emergence of life. It has been superseded—for the moment at least—by hydrothermal vents on the seabed. These vents eject heated, mineral-rich water into the deep ocean like hot springs on land. Here, hot gases and simple organic molecules might have mixed and combined in darkness to synthesize the first amino acids and other precursor compounds of life. The darkness is believed to be important because strong ultraviolet light from the sun is known to break up such delicate molecules. The Archaean ocean that covered most of the earth four billion years ago was rich in iron and other metals that might have catalyzed these early reactions, doing the job now done by the biological molecules called enzymes.

Nevertheless, the pull of the moon remains strong, and it is still easy to find scientists who cling to the idea of an origin somewhere on the tidal shores of that great ocean. Even dry textbooks occasionally betray this sentiment. David Pugh and Philip Woodworth's 2014 *Sea-Level Science* is one. "It is highly likely that we would not be here without the Moon and the tide," they write.

Part of the reason for this favoritism may lie with our deep psychology and the fact that water is central in so many creation myths. But it is one thing to say that life emerged on the edges of the sea, quite another to say that it was necessary for that sea to be tidal or that there had to be a moon in the sky.

In 1952, a graduate chemist named Stanley Miller and his supervising professor, Harold Urey, at the University of Chicago designed an experiment to simulate Darwin's pond. They combined a mixture of ammonia, methane, hydrogen, and water vapor—the obvious gaseous ingredients of amino acids—and provided an energetic stimulus in the form of heat and an electrical discharge in imitation of the lightning thought to have been a feature of the violent weather on the early earth. After a day of this battering, the water in their flask turned pink. By the end of the week, it was red and turbid with organic compounds, including three of the simplest amino acids.

Miller and Urey may have made "primordial soup" in 1952, but they did not consider the soup bowl. Was the crucible for this reaction supposed to be the atmosphere (they vaporized the water and mixed it with the other gases) or the turbulent ocean (the amino acids were found afterward in the liquid condensate)? In the latter case, shouldn't the soup have been salty? Many say our earliest environment must have been rich in salts because the organisms that evolved from it can tolerate and often require high levels of salts. Yet Miller and Urey's experiment did not include the salts and minerals known to be in the sea. Might it then have been in calm tide pools warmed by the sun that these essential chemicals were able first to concentrate before being decanted at regular intervals into the protective vastness of the ocean?

Before we get carried away with this cozy picture, it is necessary to recall that the tides when the molecular precursors of life developed were very different from those we experience today. When the earth-moon system was formed around 4.5 billion years ago, the moon was held at first in the closest orbit it could be in without collapsing into the earth only a few thousand kilometers away. A little later—that is to say, a few hundred million years later—when the earth's water first condensed into rains that themselves must have lasted for millions of years, the moon's orbit was still extremely close. The first tides would have been enormous and violent. It is in this challenging environment, riven not only by these tides, but also by volcanic eruptions and savage storms, that the first amino acids and then the complex proteins must have formed.

Aside from raising huge tides, the strong gravitational field of the moon served to stabilize the earth on its axis, which was an important factor in making life sustainable by limiting extremes of climate.

Even four or five hundred million years ago, long after the Cambrian explosion of life on earth, the moon was still significantly nearer than it is today and the tides still very high. With more than four hundred days in the year, each day was a little shorter than it is now, leaving these big tides less time to ebb and flow, so giving rise to powerful currents.

The scale of these early tides can be tentatively estimated by examining ancient rocks called tidalites, or tidal rhythmites, in which sediment layers have been built up and shaped by the regular ebb and flow of the water, although such features are thin on the ground, since they must have been raised above sea level by subsequent geological events in order not to be covered or eroded by the action of later tides. The oldest known deposit, created by tides around 750 million years ago, lies in Big Cottonwood Canyon, Utah, in which the varying force of the tides is indicated by alternating light-colored sandy layers left by strong tides able to transport sand and darker layers of mud deposited by weaker tides. The tides in this Precambrian Era may have had a range of 50 meters or more and would have run in and out at huge speeds. Beach vacations would have been challenging, but the tide pools very numerous and lively.

After his headline-grabbing synthesis of amino acids, Stanley Miller unsurprisingly made the chemistry of prebiological molecules his life's work. Scientists' ideas concerning the mix of gases and minerals available for this synthesis changed many times, but each time, Miller was able to create amino acids and other simple organic compounds. This is the easy part. Even if Miller's experiments were faithful simulations of what happened billions of years ago, nobody has been able to replicate the equally vital later stages of the process, knitting together those amino acids and other compounds to make proteins, RNA, and DNA. If anything, though, it is more likely that it is these later, more complex syntheses that require tide pools for their success. In a paper published in 1990, Miller and his coauthors speculated that DNA may first have been encapsulated in the protective fatty layers called liposomes in the intertidal zone, as also happens today. Even if this was the case, however, there is no explanation yet of how DNA was made in the first place. Was this stage, as well as the preceding one of building the proteins, accomplished in tide pools?

The answers to these questions will not only inform our knowledge of how life on earth began. They may also narrow the search for life on other planets. If a moon large enough to raise significant tides is necessary for evolution to take place—in addition to oceans and the right mix of chemicals—then the number of candidate planets we should be looking at will be greatly reduced.

11

INTO THE
MAELSTROM

A PLACE IN FICTION

In Jules Verne's classic adventure *Twenty Thousand Leagues Under the Sea*, the submarine *Nautilus*, commanded by the mysterious Captain Nemo, is careering aimlessly around the northern wastes of the Atlantic Ocean toward the end of its long undersea voyage, and Nemo's captive crew is preparing at last to leave the vessel for a quieter life on land, when disaster strikes.

> The Maëlstrom! Could a more dreadful word in a more dreadful situation have sounded in our ears! We were then upon the dangerous coast of Norway. Was the *Nautilus* being drawn into this gulf at the moment our boat was going to leave its sides? We knew that at the tide the pent-up waters between the islands of Ferroe and Lofoten rush with irresistible violence, forming a whirlpool from which no vessel ever escapes.

Verne describes a huge watery vortex,

> justly called the "Navel of the Ocean," whose power of attraction extends to a distance of twelve miles. There, not

only vessels, but whales, are sacrificed, as well as white bears from the northern regions.

It is thither that the *Nautilus*, voluntarily or involuntarily, had been run by the captain.

Jules Verne was inspired to include the maelstrom in his adventure after reading Edgar Allan Poe's short story "A Descent into the Maelström." (Both writers clearly agree that an accent is needed to lend a flavor of the exotic to the word, even if its positioning seems somewhat arbitrary.) In Poe's tale, a man is guided to a vantage point on cliffs overlooking the site of the whirlpool "close upon the Norwegian coast—in the sixty-eighth degree of latitude—in the great province of Nordland—in the dreary district of Lofoden." He sits for a few moments and then sees the sea "lashed into ungovernable fury . . . the vast bed of the waters, seamed and scarred into a thousand conflicting channels, burst suddenly into phrensied convulsion—heaving, boiling, hissing—gyrating in gigantic and innumerable vortices, and all whirling and plunging on to the eastward with a rapidity which water never elsewhere assumes except in precipitous descents."

Then, terrifyingly, the disturbance is quelled by a still greater force, and the waters reassemble themselves in a single circular current "more than half a mile in diameter" with an outer rim of "gleaming spray," inside which lies "the mouth of the terrific funnel, whose interior, as far as the eye could fathom it, was a smooth, shining, and jet-black wall of water, inclined to the horizon at an angle of some forty-five degrees, speeding dizzily round and round with a swaying and sweltering motion, and sending forth an appalling voice, half shriek, half roar, such as not even the mighty cataract of Niagara ever lifts up in its agony to Heaven."

After the two men have paused for a while to admire the phenomenon from the safety of their aerie, the guide launches into a mariner's tale of a time when a storm swept his small ship into the vortex, where it was spun around remorselessly with the splinters of other unlucky vessels, being sucked deeper with every turn. He lives to tell the tale by clinging to a buoyant cask while his ship is dragged down to the ocean bed and broken against the rocks. His first impulse, though, he says, on finding his situation hopeless, was to embrace death in a kind of ecstasy, comprehending the terror of the Maelstrom as a great manifestation of beauty. The narrator's aerial viewpoint allows Poe to heighten the reader's sensation, contrasting a familiar vertigo—frightening enough—with the unseen and unknown terror of the deep, following Edmund Burke, who, in his 1757 essay on the "sublime" in nature, pronounces that "height is less grand than depth."

Although it teems with factual geographical detail, Poe's story is usually collected with others as an early example of science fiction, and might easily be thought to be entirely imagined. The "Norway Maelstrom" is mentioned briefly in *Moby-Dick*, but this too can hardly be taken as evidence of its reality, while the fact that there was until recently a Maelstrom ride—"A High Seas Norwegian Adventure"—at the Walt Disney Epcot theme park in Florida would seem to consign the phenomenon forever to the realms of fantasy. Certainly, I had no reason to think otherwise. Like most of us, I imagine, I had thought of the maelstrom in the vaguest terms, as some kind of terrible storm, and used the word myself only in the conventional metaphorical way to denote a state of turbulence quite severed from any specific geography.

To my shame, I did not attempt either of these famous stories until another narrative had awakened my curiosity about this fabled effect of nature. Then I read A. S. Byatt's tricky little novel about the folly of biography, *The Biographer's Tale*. The central character in this delicious fiction is ambitious to advance in academia and thinks the way to do it is to write a biography—a biography, just to make matters difficult, of a biographer, a biographer who, we duly learn, seems in turn to have begun work on the lives of three people: the Norwegian dramatist Henrik Ibsen, the Victorian scientific polymath Francis Galton, and the Swedish naturalist Carl Linnaeus. The reader gathers unreliably—through Biographer One's discovery of Biographer Two's fragmentary notes—that Linnaeus actually went to see the Maelstrøm. (Yes, Byatt, too, is careful to employ a diacritic, though one different from that of Verne and Poe.)

But this was still a novel, I reminded myself. Did the maelstrom exist or didn't it? Did Linnaeus once go there? Does it exist still? It is impossible to unravel fact from fiction, or even one fiction from another. Rereading Byatt's book now, I find that Biographer Two's "notes" rely heavily on Poe's description of the maelstrom, quoting verbatim the passage I gave above, with the sole difference that the whirlpool has crept up in size from half a mile to "more than a mile" in diameter. Byatt cunningly reveals at the end of her story that the principal character (Biographer One) has become aware of a deceit perpetrated by his would-be biographical subject (Biographer Two), who is guilty of recycling Poe's words. Biographer One has learned the hard way that the surviving "facts" of a lived life are never complete or stable. And his subject? Biographer Two, he learns, seems to have perished in the Maelstrøm during the course of his overenthusiastic researches. He has gleaned this last piece of information from a newspaper cutting reporting the man's demise, which offers this laconic aside: "The search for authenticity in scholarship can have its dangers."

THE MAELSTROM does sound like a fiction—a supernatural force with the power to carry men off to their death, located conveniently far from any reader's direct experience. It turns out that Linnaeus truly did go in search of it on a tour to Lapland in the summer of 1732, but even he was thwarted when he could not find a boat to take him the final distance.

Although the writers I have mentioned drew on articles in contemporary magazines, as well as on each other, for their inspiration, the maelstrom's rise to prominence in fiction may be ultimately attributed to the arguably more factual account of Erik Pontoppidan, the Danish-born bishop of Bergen, who found time away from episcopal duties to publish what he claimed to be the first *Natural History of Norway* in 1752. His work followed the example of Linnaeus in offering a detailed scientific survey of the people, animals, and plants of the country, and included a discussion of the tides and their causes, for which the bishop relied on classical authors, as well as the newer ideas of Descartes. "There is," the bishop wrote,

> another kind of current, or motion of the water in the sea of Norway, remarkable, and somewhat relative to the ebb and flood, named the Malestrom, or Moskoestrom, in the 68th degree, in the province of Nordland, and the district of Lofoden, and near the island Moskoe, from which the current takes its name. Its violence and roaring exceed those of a cataract, being heard to a great distance, and without any intermission, except a quarter every sixth hour, that is, at the turn of high and low water, when its impetuosity seems at a stand, which short interval is the only time the fishermen can venture in: but this motion soon returns, and, however, calm the sea may be, gradually increases with such a draught and vortex as absorb whatever comes within their sphere of action, and keep it under water for some hours, when the fragments, shivered by the rocks, appear again.

Like many authors of early natural histories, Pontoppidan was remarkably scientific in his intent and his methods. Nevertheless, Bergen is a long way south of the Nordland, and it is unlikely that he saw for himself many of the things he describes. He manages to blur the distinction between its authentic fauna and figments of the imagination such as sea serpents and mermaids by taking the testimony of unreliable witnesses at face value. The kraken also entered into the English popular imagination through translation of Pontoppidan's work, though it is found in older natural histories (such as the Swedish Olaus Magnus's *History of the Northern Peoples* of 1555) and in Norse sagas too. Does the kraken exist? In a way, it certainly does. Descriptions of the creature correspond closely enough to the giant squid, and it is known from sucker marks found on whales that specimens of giant squid exist far larger than have yet been seen.

And in that case, can we believe the bishop's description of the maelstrom? Perhaps we can, for on this occasion he was probably taking the word of a trustworthy local source. This was the seventeenth-century Nordland poet and clergyman, Petter Dass, whose verse work, *The Trumpet of Nordland*, documented much of the natural history of his region. Among his observations of the maelstrom, Dass detected a correlation between the strength of the current and the phase of the moon, and reasoned that the great speed of the vortex was a consequence of the tidal flow in the area having to squeeze through a narrow channel between islands.

Using Dass's observations, Pontoppidan felt confident in refuting a belief popularized by the German Jesuit Athanasius Kircher, a scholar in geology and other sciences, that the maelstrom must be the entrance of a subterranean channel passing under the Scandinavian peninsula—a kind of giant plughole linking the Atlantic Ocean to the Gulf of Bothnia. He also offers the thought that the strait containing the maelstrom, between the tip of the southernmost of the main Lofoten islands and the tiny rock of Mosken, might even be the Scylla and Charybdis of classical mythology. If this is so, then Odysseus must have drifted far, indeed, from his course home across the Mediterranean Sea to Ithaca.

The word "maelstrom" presents another puzzle. For a start, I cannot get to the bottom of the accent question. I have seen the first syllable given as Mahl-, Mael-, Mål- and Mal-, and the second as -strøm, -ström, -straum, -straumen, and -strom ; you may take your pick. The *Oxford English Dictionary* gives several more permutations and cites as the first usage of the word in English a 1589 account of the voyages of the explorer Anthony Jenkinson, who made several expeditions to Muscovy, forging trade deals between Elizabeth I's England and Ivan the Terrible's Russia. "There is," Jenkinson wrote, "between the said Rost Islands, and Lofoote, a whirle poole, called Malestrand [*sic*], which . . . maketh such a terrible noise, that it shaketh the rings in the doores of the inhabitants houses of the said Islands, ten miles of." But Jenkinson is thought to have got the name off Dutch maps, not Norwegian lips. In Dutch, the word means, roughly, "grinding stream," and the word for this—and now any—marine whirlpool in all Scandinavian languages stems from this original.

PRINCE BREACKAN'S CAULDRON

On the other hand, I was confident enough of the existence of Britain's most notorious whirlpool. This is the Corryvreckan, which lies between the islands of Jura and Scarba in the Inner Hebrides off the west coast of Scotland. As with the Maelstrom, even its name is the subject of some confusion. In Gaelic, *coire* means "cauldron." *Breacan* is the word for a more or less checkered pattern and is used to describe mackerel skies, plaid cloth, and the plumage of certain birds. It might also describe a striking pattern of waves on water, such as that kicked up by a tidal race. Corryvreckan might therefore be translated as the "cauldron of the choppy sea." But it might equally be the "cauldron of the plaid," an appellation perhaps acquired from Gaelic mythology in which this whirlpool is the washtub of the creatrix Cailleach, where she is supposed to wash out her woollens in preparation for the winter.

A different derivation of the name heard locally is recounted in Michael Powell and Emeric Pressburger's superb allegorical film of 1945, *I Know Where I'm Going!* Headstrong young Joan Webster has set out from Manchester for the remote Scottish island of Kiloran, which is owned by the wealthy industrialist she intends to marry. Waiting for the storm to die down that prevents her from making the final stage of her journey, she observes a picture on the wall. It depicts the Norwegian prince Breackan, who came across the sea seeking the hand in marriage of a princess of the isles. A handsome naval officer on leave, Torquil MacNeil, who is also waiting to get to the island, tells her the story: the princess's father will agree to the marriage if Breackan can demonstrate his seamanship by riding at anchor for three days and nights in the whirlpool. The prince returns to Norway to seek the advice of his elders, who tell him he must take with him three anchor ropes—one made of hemp, one of flax, and one woven from the hair of maidens faithful to their lovers. He returns to Scotland duly prepared.

Inevitably, the first two ropes break. The third rope— But at this point, the telling is interrupted by the impetuous Joan, who, exasperated by the continuing delay, pays a local lad an absurd amount of money to take her out to the island in his fishing boat regardless of the storm. MacNeil heroically leaps aboard just as the little vessel is setting off. The boat runs close to the whirlpool on its way to the island, and then the engine dies. Gradually, they are sucked toward the vortex, which is portrayed in remarkable montage sequences of great technical ingenuity. These, as Powell explained in his memoirs, combined live footage taken using a handheld camera in a small boat, "mostly shot by me tied to the mast like Prince Brecan," with sequences filmed using high-speed cameras and a model boat in a studio water tank with gelatin mixed into the water to alter its viscosity in order to give the seas the right majestic scale—tricks Powell had picked up from the parting of the Red Sea in Cecil B. DeMille's silent epic *The Ten Commandments*. But, as Powell wrote, the "director of the sequence was really Edgar Allan Poe."

In the midst of the chaos, MacNeil—who we have learned by now is the true laird of the unreachable Kiloran—reveals the final installment of the story: the rope made of maidens' hair holds until the turn of the tide. But one maiden—only one—had been untrue, and when that strand broke, so did the rope. Finally, MacNeil manages to restart the engine and they escape the whirlpool.

Breackan of the legend is not so lucky. He is drowned, and his body is washed up in a nearby cave, where he is supposed to lie buried to this day. As for Joan, she is forced to accept that chance must play a role in her life, having naturally fallen for the navy man. Powell and Pressburger's very British films are usually laden with very un-British expressionist symbolism. In this film, it is the spiraling whirlpool that has made Joan see that she does not, after all, know where she is going.

PERHAPS GEORGE ORWELL never saw the film, for a year or so after it came out he was nearly drowned in the Corryvreckan, along with his young son, Ricky, and a niece and a nephew, Henry and Lucy Dakin, whom he had taken out in a small boat while staying on Jura to work on what he knew would be his last novel, *Nineteen Eighty-Four*. Orwell was no mariner, and it seems he may have ignored the advice of others who knew more about the local conditions.

The Admiralty Pilot for this stretch of the British coast at the time gave fair warning: the Corryvreckan, it said, is "noted for its turbulent waters," which run at up to 8½ knots at spring tides. When the tide is flooding, it produces eddies, "causing breakers, which are sometimes so heavy as to be dangerous to small vessels." In general, "navigation at times is very dangerous, and no stranger would be justified in attempting it." It is seldom that one can detect any note of emotion amid the laconic drone of information in these essential guides to navigation, but in this entry there seems almost to be an edge of suppressed hysteria.

Orwell was doubtless not armed with this volume, and perhaps was fooled by the fact that the *vertical* range of tides in the area is very low; there is an amphidromic point where there is no tide at all nearby in the Sound of Jura. So it was that his party set off somewhat earlier than advisable, before the tide had slackened sufficiently for the whirlpool to abate and the waters to be safe for passage. As their outboard motor struggled to push the little boat onward, they found that they were making little progress relative to the land, and that the waves were getting gradually bigger. Soon they realized they were in the outer edges of the Corryvreckan—not yet in the powerful funnel of the central whirlpool, but tossed between the smaller eddies swirling around it. Orwell struggled to hold a course with the tiller, but then the outboard broke from its bracket and sank into the sea. They were now entirely at the mercy of the tide. Indicating his own weak chest, Orwell directed his nephew to start rowing. Henry struggled to pull the boat clear of the whirlpools but could not do it. However, he was just able to hold their position until the whirlpools gradually died away with the continuing ebb tide. Eventually, the water's grip lessened, and he was able to row clear and ferry the family to a nearby island in the stream.

The misadventure was not yet over, though. While they were attempting to land on the island, the boat flipped over in the Atlantic swell. Orwell and his three young companions barely scrambled ashore. Most of their possessions were lost along with the boat. They were lucky to be rescued by a passing lobster boat only a couple of hours later.

What is oddest about the incident, in retrospect, is Orwell's attitude. Even though members of his family were endangered, his mental state seems to have been one of extreme detachment. While the boat swirled in the current, he interjected irrelevant nature notes, as when a seal popped its head out of the water near the whirlpool, for example. Perhaps it was an attempt to maintain the illusion that it was all an adventure. Ricky, aged three, seems not to have been afraid, recalling only the "up down, splish splosh" of the ride. Henry Dakin, however, afterward reported that Orwell had confided in him, while the others were out of earshot, that "he thought we'd had it." An earlier visitor on Jura had found the author in "a sort of self-destructive mood." Was Orwell then in the grip of a death wish, as somebody later put it to Henry? "I said I hardly thought so; if he'd had a death-wish he'd hardly have taken three other people with him. He wasn't that kind of chap at all."

ASTONISHINGLY, it was only in 2012 that a team of oceanographers, led by John Howe from the Scottish Marine Institute, resolved the mystery of the whirlpool. While making a detailed hydrographic survey of Irish and Scottish waters, they identified a long undersea cliff off the shore of the island of Scarba. The bathymetric survey, made using a variety of echo-sounding techniques, also revealed a seabed scoured out by the strong tides to expose the bedrock, with all the mud and sand piled up in underwater dunes where the currents finally relent.

The present survey was undertaken in order to furnish reliable data for marine conservation projects and potential renewable-energy schemes, but also to improve the safety of navigation. The British Admiralty's existing chart, *Gulf of Corryvreckan and Approaches*, gives isolated depth soundings here and there as on all charts, but some of the data are based on readings taken by lead line, perhaps dating back as far as the nineteenth century, and the chart does not provide the detailed contours that would reveal the presence of such a feature. Nor does it give any clue to indicate the presence of the isolated pinnacle of rock in midchannel that many had presumed to be the cause of the whirlpool, the existence of which had supposedly been confirmed by a previous survey. "The pinnacle, the pinnacle, everybody was saying. I was looking for this, and the old survey data had this," John tells me. "But when we surveyed it, the rock is a wall, or a series of walls and ledges, sticking out from the Scarba shore." No pinnacle. Instead, a series of rock buttresses that stretches along the Scarba shore. One buttress reaches to less than 30 meters from the sea surface. It is here, when the tide is flooding into the Sound of Jura, that large volumes of water well up over the cliff and are forced to run faster through the shallower water across the top. When the tide ebbs, the direction of flow is reversed, and the water pours over the cliff edge and is suddenly pitched into the deeper sea, causing water behind it to be sucked into a downward swirl, the Corryvreckan.

NORTH

In fact, whirlpools are a rare but dramatic and genuinely life-threatening feature along many seacoasts. The Corryvreckan is reckoned to be the third largest in the world. The largest in the Western Hemisphere is the Old Sow in Passamaquoddy Bay in the outer Bay of Fundy between Maine and New Brunswick. Captain FitzRoy got caught in one off the coast of Patagonia in 1833: "an irregular motion in every direction, exactly like the boiling of a pot, on a great scale." FitzRoy plumbs its depth and finds no bottom—a frequent characteristic of such places. But the largest whirlpool of all is the Maelstrom, or Moskenstraumen as it is locally called, and I need to see it.

The stories left both the maelstrom (natural phenomenon) and the Maelstrom (named geographical location, accents optional) still poised between reality and unreality, fact and fable. I was sure there would be something there. But what? Would it be as terrible as the vision conjured by Poe and Verne? I had avoided Charybdis for fear of being underwhelmed by the actuality in comparison with the classical version. Was it perhaps true that, as one of my textbooks on tides laconically put it, the Maelstrom and Charybdis are "more celebrated in literature than they are in oceanography"? And if so, was this a fair indictment? Hilaire Belloc, too, reckoned that the tidal race off Portland Bill could "eat" the Maelstrom "and not know it had had breakfast," although he was honest enough to admit that he hadn't actually been there. What did these casual dismissals say about our changing attitude to the natural world? What did they say about the power of stories? If the Maelstrom existed, it would be conclusive proof of the power of the tide as a maker of places. If it did not—well then, hats off to the storytellers.

It will take me three flights and a ferry journey to reach the Lofoten Islands in the Nordland province of Norway. To my soft, southern ears, it sounds like the north of the north.

Like most British people, my idea of the marine north is daily imprinted on my brain by the radio shipping forecast:

> . . . issued by the Met Office on behalf of the Marine and Coastguard Agency at 0505 on Thursday. . . . There are warnings of gales in Fair Isle, Faeroes and Southeast Iceland. The general synopsis . . .

The names of the "sea areas" into which the waters around the British Isles are divided are inspired mostly by the adjacent terrestrial geography, but two evoke naval glory: Trafalgar, and FitzRoy, the whirlpool survivor who later introduced the first telegraph weather warnings for shipping. The litany begins and ends in the north. Sea area Viking is always the first to be read out. The last is

> . . . Southeast Iceland. Northwest gale force nine. Snow showers. Moderate, locally poor. No icing.

I remember listening as a small child to that habitual final cadence, "No icing," and thinking what an unhappy place Southeast Iceland must be. No icing. Only later did I learn that icing is a thing much to be feared, the encrustation of a ship's rigging with layer upon layer of black ice until its top-heavy weight can be enough to induce a capsize.

There have been only a few alterations to the sea areas over the years. In order to harmonize with the equivalent Norwegian scheme, the part of Viking nearest the Norway coast has been subdivided into two new sea areas:

> North Utsire, South Utsire. Northwest six. Fair. Good.

How exotic these places sound, certainly compared with Humber and Thames, my local sea areas. They lie often an entire weather system away. While we bask in windless sunshine, they are lashed by storms. They sound remote indeed, but I find they are not, in fact, all that far north. The names are taken from the small island of Utsira, which lies just off the coast near Stavanger in the south of Norway. Even North Utsire is far to the south of where I am heading.

I could draw a creditable map of the Mediterranean, and I expect you could too. The Romans called it *Mare Nostrum*, "Our Sea," and it has become our sea as well. But the outline of the closer coasts of the North Sea north of the British mainland is lost to our memory, although it was once of vital importance. Where I am going is practically off the edge of the map now, nearly 68 degrees north, about 130 miles inside the Arctic Circle, north of the whole island of Iceland, north of almost the entire state of Alaska.

The logistics of the journey are only one part of it. As with my other tidal rendezvous, I have to time my trip carefully to coincide with suitable tides. I can do nothing about the weather, which for all I know might also have to be ideal in some respect, but I can at least use the tide tables to ensure that I will see the Maelstrom at a time when the tides are running at their most furious.

Perhaps then I will avoid the disappointment that Edmund Gosse suffered when he traveled through Norway in 1871 and "discovered" the playwright Henrik Ibsen for the English-speaking world. In preparing for this trip, I was disconcerted to find that the noted writer and critic set out, like me, by congratulating himself for resisting the southward drift of the common traveler, and heading north instead to the "noblest coast scenery in Europe." He duly arrived in the Lofoten Islands—"associated with little except school-book legends of the Maelstrom, and perhaps the undesirable savour of cod-liver oil"—and found the scattered population of fishermen busy bringing in the spring catch of cod.

He sought out the Maelstrom, yet he did not find it and, with the

confidence that perhaps only a Victorian critic can muster, took his failure to mean that the thing did not exist:

> It is a matter of regret to me, in my functions of apologist for these islands, that truth obliges me to raze to the ground with ruthless hand the romantic fabric of fable that has surrounded one of them from time immemorial. The Maelström, the terrific whirlpool that "Whirled to death the roaring whale," that sucked the largest ships into its monstrous vortex, and thundered so loudly that, as Purchas tells us in his veracious Pilgrimage, the rings on the doors of houses ten miles off shook at the sound of it this wonder of the world must, alas ! retire to that limbo where the myths of old credulity gather, in a motley and fantastic array. There is no such whirlpool as Pontoppidan and Purchas describe.

Gosse went on to describe the location of the whirlpool accurately enough, and gave its Lofoten name. He recounted seeing fjords where the tide running against the wind whips up the water in a way that he can imagine might overwhelm a small fishing vessel, so it is clear that he knew what he was looking for. What went wrong? I can only assume that he went at a state of the tide when the waters were still. I must be careful not to do the same.

Oddly, although he has refuted the existence of the Maelstrom in such emphatic terms, Gosse is happy to celebrate its constancy in literature. Indeed, he contributes his own obscure source to the growing list of those who have described it, in the person of Anders Arrebo, a Danish-born bishop of Trondheim in the early seventeenth century and, he adds, "the father of modern Scandinavian poetry." Arrebo's *Hexameron* is an epic-verse rendition of the six days of the biblical creation relocated amid the crags and fjords of Norway. His Nordic Eden is not complete without the Maelstrom, on which his lines run (in Gosse's translation):

In Loufod far to north on Norway's distant shore,
A flood is found that hath no like the wide world o'er,
Entitled Moske-flood, from that high Mosker rock
Round which in seemly rings the obsequious waters flock;
When this with hasty zeal performs the moon's designs,
If any man comes near, the world he straight resigns;
In spring its billows rear like other mountains high,
But through their sides we see the sun, the earth's bright eye;
Then, if the wind should rise against the flood's wild way,
Two heroes rush and meet in crash of war's array;
Then tremble land and house, then doors and windows rattle,
The earth is fain to cleave before that monstrous battle;
The vast and magic whale dares not its breach essay,
But turns in fear to flight, and roaring speeds away.

I feel Arrebo's description has an authentic smack. It refers to a time of year when the phenomenon might be most active and to its dependence on the state of the tide ("the moon's designs"). True, there are none of the wild vortices of Verne and Poe, but the impression of terrible power is clear enough. Surely this Maelstrom existed, even if later ones didn't. But there is another possibility I can't rule out. Gosse may be dissembling, pretending that the Maelstrom no longer exists precisely in order to buff the reputations of the literary figures he is championing as fabulists.

BRIDGE OVER TROUBLED WATER

My final flight deposits me in Bodø, on the eastern shore of the Vestfjorden, a triangular maw of water open to every southwesterly gale. From here, it is a three-hour ferry journey across the mouth of the fjord to the islands. The place is not much, but it is clearly a great metropolis compared to where I am headed. Gosse wrote that the

little town is "London and Liverpool in one for the inhabitants of our islands: every luxury, from a watch to a piano, from a box of Huntley and Palmer's biscuits to a pig, must be brought from Bodo."

Before I embark, though, I have time to whet my appetite with a visit to another site of tidal turbulence, although this one has garnered no stories, only one bald statistic. For, a little way down the mainland coast from Bodø is the tight entrance to a fjord, where the world's fastest tidal current runs at speeds of up to 20 knots. Almost as fast is the tide that funnels through the Naruto Strait near Tokushima in Japan, depicted in one of the last works of the famous master of *ukiyo-e* colored woodblock printmaking, Utagawa Hiroshige, and in beautiful gold-leafed panels of the Edo period. But as far as I know, there are no historic images of the Saltstraumen.

The gulls have assembled on the lampposts of the bridge over the stream when I arrive at half tide. I lean over the railing and look down at the water. From this height, the influx is already dizzying to view for more than a few seconds. It is more like being on the bridge of a ship than on a fixed structure. The current runs fast and smooth in midchannel and throws up boils at the edges of its self-made stream, where water deflected from the bottom and sides of the inlet finds it has nowhere to go but up. Periodically, rows of white overfalls are scuffed up as the water trips over itself in its hurry to fill the fjord. An eider bobs through the narrows at high speed, enjoying the ride. It makes me laugh out loud. I have never seen a sitting duck move so fast.

Although the range of the tide hereabouts is tiny, I notice there is an unnatural slope on the water as it pushes in through the narrows. The water level on the seaward side is perhaps a meter higher than the water in the fjord. The rocks on the bank are smoothed away on top by the waves of centuries, but below, where they are more often covered, they are etched into fine runnels that remind me of whale hide.

As the tide accelerates even more, massive wracks are tossed around like mermaid's hair. Little whirlpools form and chase their tails, gulping and merging as they sweep upstream. The water is now pushing in so fast that the main influx can no longer hold a straight line, but bends in powerful loops like rope pushed into a hole. Small but sharply funneled whirlpools now gather continuously on the foaming edge of the main current. Bigger whirlpools form where the line between the moving water and the still water cannot hold, and a fast stretch begins to overtake a slower one. The whirlpools curl outward from the center, rotating clockwise on the right side of the inflowing channel and counterclockwise on the left. There is none of that nonsense about plugholes and hemispheres here; the local forces are too great for that.

My reverie is abruptly spoiled by the appearance of a couple of rigid inflatable boats (RIBs) laden with tourists in high-vis oilskins. The lead driver waves to me. Jerk, I think. A little later, when the tide is high and slow, a few local fishermen appear in their workboats. I take off to kill some time by having lunch at a deserted café overlooking the stream. I am surprised to find what appears to be an abandoned swimming pool outside, its walls now cracked concrete. It was not for human swimmers, I'm told, but for seals. Through a plate-glass window in the side of their aquarium, the animals would have been able to see the rushing tide far below. I am not sure this is a kindness.

I return to catch the ebb. The efflux is different—smoother, blacker, more concentrated somehow, like oil poured from a barrel. At the edges of the laminar flow, the clear water is churned to the turquoise color of a mint candy. The rushing sound it makes is loud and constant. A breaker appears, like that on a coral reef, and grows to reach almost the whole way across the channel, even though there is no reef; it is only the press of water rushing to escape from the fjord.

So much energy—and the whole drama will be reenacted in a few hours' time. And ever after.

THE NAVEL OF THE OCEAN

SØRVÅGEN, NORWAY

	HW	LW	HW	LW
June 15, 2014	01:51	08:22	14:22	20:35
	3.0 m	0.3 m	2.8 m	0.3 m

Crossing the Vestfjorden on a large ferry, I see small black auks flying under our bows. An announcement alerts passengers to a pod of killer whales off the port beam. But I see no other ships.

Soon the Lofotens come into view through the salt-stained windows. Steepening mountains rise to bare, sharp peaks all along the spine of the islands, forming the Lofotveggen, or Lofoten Wall, which serves as brute protection against the north winds off the Arctic Ocean. Most of the villages are located on the lee side of this natural barrier, facing onto the Vestfjorden, where they also catch the sun. The little red huts cluster agreeably on stilts over the sea, some of them still with their traditional turf roofs. Formerly, they were rented to fishermen who would sail up from the south of Norway each year for the summer cod harvest. Now, they are used by tourists, although enough cod is still caught here that a sweet, ammoniacal smell fills the air from the timber racks driven into the rocks where the fish is hung out to dry before it is exported.

I make my way to the southernmost of these villages, closest to the tip of the Lofoten Islands, though still 4 or 5 miles from the Moskenstraumen, and consider how best to accomplish the final stage of my journey. Although it is not far, I am disappointed to learn that it would be virtually impossible for me to continue on foot to Edgar Allan Poe's fictional vantage point over the Maelstrom on the slopes of Helseggen Mountain (or "Hellsegga," as Poe called it). Looking south from a convenient promontory, it is easy to see why this is so.

Sheer, gray sheets of three-billion-year-old rock plunge straight into the fjord for as far along the coast as I can see. Below the surface of the cold, blue sea, they presumably continue their dive into the deep.

I have scorned thrill seekers who want to lord it over nature by crashing through the tide accompanied by the roar of engines, but now I find my only option is to join a boat trip if I am to see the Maelstrom. It is late afternoon, and although the sun will not set tonight, its rays are already blocked by the Lofoten Wall as we ease our way out of the pretty harbor. The journey in the twelve-seat RIB feels like a theme park ride as the fabulous scenery shudders by frame by frame. The aged slopes lour in the shade as sea eagles wheel lazily above us. Occasionally, we pass a deserted fishing village. Settlements here date from at least the seventeenth century, but many of them were abandoned after the Second World War, when Lofoten island-ers were offered incentives to take up occupations in industry on the mainland as part of a national modernization plan.

Others had left before, often for the Great Lakes states in Amer-ica, driven out by the dangers of the fisherman's life and the iniquities of the serfdom under which they had to work. Their story is told in Johan Boher's novel *The Last of the Vikings*, a Nordic version of *The Grapes of Wrath*, which traces the long westward journey in pursuit of an easier life on a distant continent. Boher's Norwegian scenes also make it plain that, to these brave men, the Maelstrom was simply one of the many hazards of their trade that had to be negotiated routinely in order to reach the Atlantic fishing grounds. For them, the Lofoten adventure lay in the promise of the riches they might make in a good cod year.

A red-capped light on a black-and-white pole marks the southern tip of the main Lofoten Islands. Only the smaller islands of Vaerøy and Røst loom ahead, and before them Mosken and the numerous skerries of the Moskenstraumen passage.

It is approximately four hours after high water, and I reckon the tide should still be ebbing out of the Vestfjorden through the gaps between the islands into the ocean beyond. Lars, who is driving the boat, says he has seen waves 20 feet high here, churned up by the ocean swell and storm winds as much as by the tide. But today the sea is calm. The surface of the water is ruffled by small waves, but they ride happily on the broader swell and do not break. As we surge onward, here and there appear what seem to be separate pools of water, smoother and glassier than the rest, mysteriously unruffled by the waves. These are boils, massive upwellings of water from deep below, whose outward spread when they reach the surface irons out the wind-made waves. Their smoothness is not the smoothness of the proverbial millpond, but of something being actively made, like a surface of molten metal.

Small waves break at the edges of these glossy patches like ripples on a beach. It is as if the two waters are too different to mix. Perhaps the upwelling water is colder or saltier than the water close to the surface. At the edges of the boils, too, are little eddies that occasionally intensify into downward whorls as the thrust-up water finally runs out of momentum and begins to sink again.

They do not last long, however, and soon the eye is drawn to a new mirror sheet somewhere ahead or on the other side of the boat. They form continuously over a wide expanse of the water, some of them several hundred meters across. Lars slows the RIB to idling speed so that we can feel the boat move beneath us. It is an unfamiliar sensation. We are not wallowing in waves, but are borne up and twisted this way and that like a cork in a fountain. Yet there are no obvious corresponding vortices where the water is sucked down, and there is certainly no giant, ship-endangering whirlpool.

We motor on to a landing site on the ocean side of the islands—the sunny side of the Lofoten Wall in this evening of a midnight-sun day—where there is a cathedral-like cave in the mountainside containing Bronze Age wall paintings of human figures. Northern Norway is of special interest to archaeologists because it offers evidence of coastal settlements from as early as the Neolithic period, which have been raised above sea level as the land has rebounded with the weight

of ice-age glaciers no longer pressing down on it, whereas similar settlements further south have been lost beneath the sea.

Close to the coarse coral beach where we disembark, stacks of
bleached wood lie stranded above the tide line looking like one of
Andy Goldsworthy's tidal artworks in the later stages of decomposi-
tion. It is an odd sight, given that no trees grow around here. However,
Christian, our Lofoten-born guide, tells us about a Russian cargo ship
that got into difficulties in the Moskenstraumen a few years before
and was forced to jettison its load of timber in order to make the vessel
safe in heavy seas. It seems the Maelstrom has not quite lost its fabled
power. The story echoes the fishermen's lore that one should throw an
oar or other large timber into a whirlpool in order to escape with one's
vessel and one's life.

Here and there are also large logs that have washed up. Christian explains that they probably escaped retrieval while being floated
downriver during the course of logging operations in Siberia, and
were then transported by currents around the Arctic Ocean. Such
timber flotsam was once an invaluable resource for those living on
these bleak, treeless islands, and it was eagerly snapped up for houses
and boats. Sadly, I also see much else that does not fit in as well—the
usual fluorescent nets and plastic containers and other paraphernalia
of modern fishing and other offshore industries, as well as a surprising amount of recognizable consumer waste, such as drink bottles.
All in all, there is enough waste that even in this remote region it
has become necessary to organize beach cleanup parties, and bulging refuse bags containing the collected material also sit on the shore
awaiting more suitable disposal.

An oar or a spar from the deck of a boat may not always be enough
to appease the maelstrom. Occasionally, a greater sacrifice is called for.
In a legendary story of a pirate ship caught in the Naruto whirlpools,
the pirates thoughtfully make an offering of the clothing of the princess whom they have just kidnapped. When this does not work, they

debate whether they must throw their captive into the raging waters. As they argue, however, a divine message orders them instead to build a boat that will carry the princess to freedom. It is understood that these offerings are made to appease the whirlpool deity, but they are based on good science: a suitable object may indeed break the vortex and cause the whirlpool to fill for long enough to escape its clutches.

THREE HOURS after passing through the Moskenstraumen in a westward direction, we are heading back through it into the Vestfjorden. It is now an hour after low water at Sørvågen, the nearest place for which the United Kingdom Hydrographic Office issues data, although it feels as if the tide is still ebbing.

This time there is more to see. Ahead of us, a large area of gleaming water contrasts with the choppy surface all around its edge, dark and dappled in the angled light. The texture of the sea surface is like boldly applied oil paint on a canvas. Furious little waves make energetic forays across the top of the upwelled water. They break foamily and are instantly replaced by new waves on the same line. We cross this divide—so real in oceanographic terms, a live border between different seas, and yet so inconsequential to us in our powerful boat skimming the surface at speed.

In the distance, I notice grander arcs of these standing waves—first one, made conspicuous by large white breakers, then another, concentric with the first. They seem to mark the edge of something. But the something—we are in it. I think I recognize the scene. "We were now in the belt of surf that always surrounds the whirl; and I thought, of course, that another moment would plunge us into the abyss," Poe's narrator says. But there is no downward vortex. It is not the giant whirlpool of legend. The large area of chaotic water encircled by the breaking waves is, in fact, eerily smooth.

The crests are breaking in such a way that they appear to be chas-

ing one another along the distant wave front. Perhaps it is entirely
a hydrodynamic effect, or perhaps it is an effect of the wind. Either
way, the impression is of a majestic clockwise orbit, giving the clear
sense that we are in the midst of a majestic gyre several hundred yards
in diameter. The phenomenon may be transient: when he is delivered
from his watery abyss and finds himself "on the surface of the ocean,
in full view of the shores of Lofoden," Poe's narrator marvels that he
is safely afloat "above the spot where the Moskoe-ström *had been*."
Yes, it may be transient—dependent on the right phase in the cycle of
ebb and flow, and on the tides' being strong enough (after all, I have
purposely come here during spring tides), and perhaps even on particular weather conditions—but it is also undeniably primal. Its circle
represents both chaos and order: it is a perturbation of the "normal"
movement of the sea and the imposition of a new, disturbing pattern.

It is an uncomfortable feature to be facing at water level. Were I at
the helm of a sailing vessel, I would not want to be here at all. Even
though the water is largely flat, it is definitely not calm. The energy is
palpable under the surface. The fact that the tide has the strength to
reduce the waves whipped up by the wind to nothing is a clue to the

greater might of the deep. It is a quiet, venomous authority in contrast to the petulant rage of great waves. But our boat, with its twin 250-horsepower V6 engines, powers on obliviously.

MYTH AND MATHEMATICS

Once, it was believed that this was where the earth drank and then violently spat out what it did not want. Or, it was the site of great undersea stones that ground the salt for the oceans. Of course, mariners invoked sea monsters as an explanation for the preternaturally lively sea, with its sliding currents and licking waves; such action begged animation. It is no surprise that beasts like these are prominent in Norse mythology, and especially in Norway, which, with all its islands and fjords, claims to have 100,000 kilometers of ocean coastline. After all, which is worse? Which is more of a consolation? Which is more *plausible*? That the sucking tentacles of a giant squid, the kraken, have taken a life, or that the suction of the sea unaided has done it? It must have been preferable to imagine a freakish creature as

the malign force that can drag down a ship rather than having to hold constantly in mind the knowledge that the sea alone can do this—the very sea that you are floating on, or are bound to cross.

Marginally more rational was the notion, promulgated by Bishop Arrebo, among others, that the Maelstrom marked the entrance of an underwater channel linking the ocean with the Gulf of Bothnia by passing under the Scandinavian peninsula. (Such beliefs, which seem absurd now, were once widespread. It was said, for example, that the wrecks of ships lost in the Atlantic would sometimes surface in a particular lake in Portugal, while the water in the Arethusa fountain near Siracusa on Sicily was long believed to come from the Holy Land, which is presumably a Christian adaptation of a Greek myth that it came under the sea from Arcadia.)

In 1997, a team of mathematicians from the University of Oslo decided it was time to take a more scientific approach. "In view of the early interest in the phenomena [*sic*] it is surprising that there have been no substantial modern studies of this strong tidal current," wrote Bjørn Gjevik, the professor of hydrodynamics who led the research. The objective of the study was not to pronounce on literary fancies, however, but to add to knowledge of the tidal-flow patterns, which are thought to be crucial in dispersing young cod, hatched in the nursery grounds of the Vestfjorden, into the Atlantic Ocean.

Gjevik and his colleagues designed a computer simulation of the tidal movements around the Lofoten Islands to see whether they could finally unravel the mystery of the Maelstrom. To reproduce the complex patterns of the water's constant movement, taking into account the complex seabed topography and coastal outline, posed a huge technical challenge, even using high-powered computers. However, the model they developed was found to agree well with the observed tides.

The scientists found that a sea gradient—that is to say, a difference in surface sea level between two points—of just 12 centimeters over a distance of 10 kilometers (a slope of 1:80,000) is all it takes to drive the current between the islands with sufficient force to produce the massive whirlpool. This is far less than the gradient I had seen at the entrance to the Saltfjorden, near the town of Bodø. On either side of the Moskenstraumen, however, both in the Atlantic and in the body of the Vestfjorden, the sea falls away rapidly to depths of 200 meters and more, whereas the depth in the narrow stream is less than 50 meters—well short of the 40 fathoms or "immeasurably greater" specified in Poe's tale. This constriction means that colossal volumes of water—an estimated 350,000 cubic meters (140 Olympic-size swimming pools) per second, or about a third of a sverdrup—must pass through the channel at high speed on every tide.

The mathematical modeling also predicted various surface effects of the currents, most notably a giant eddy about 6 kilometers in diameter, which revolves slowly through the channel in either direction according to the run of the tide. His computer's surprising result inclines Gjevik to excuse some of Poe's hyperbole: "The current in the area northwest and east of Mosken has a clockwise rotary character which may mistakenly have been interpreted as a manifestation of a large whirlpool by the early observers." But it is, he confesses, "only a bleak reminiscence of the monster eddy described in the old literature."

MY OWN DESCENT into the Maelstrom prompts me to wonder why it is that we appear to have lost our proper fear of the tide. It cannot be scientific explanation of the forces at work that now causes us to treat the tide with disdain; there is still much to learn. Nor is this contempt bred entirely of the familiarity that comes of so many of us living and vacationing by the coast. The dominant factor, I believe, is the hubris that comes from our confidence that we can conquer the tide with technology. Mark Twain called it more than a century ago: "Time and tide wait for no man. A pompous and self-satisfied proverb, and was true for a billion years; but in our day of electric wires and water-ballast we turn it around: Man waits not for time nor tide." His quote is now much repeated in gung-ho business manuals.

Yet recall George Orwell and his outboard motor, and the Chinese cockle pickers out in Morecambe Bay with their van. Even Poe's narrator finds himself caught out in the Maelstrom because his watch has stopped.

Now I have been through the Maelstrom, but I have been through it with the power of five hundred horses. As I saw at the Saltstraumen and on the Shubenacadie River, our machines slip lightly over the top of each vortex and boil. They easily match the power of nature, but the thrill we then feel is the thrill of that artificial force, and not the original power of nature itself. The energies involved may be equal, but the sensations produced in our bodies, and the emotions fired in our brains, are of a fundamentally different kind. We are celebrating our own petty triumph, and using it to seal ourselves off from the power of nature.

12

SIGNAL AND NOISE

SEEKING SEA LEVEL

STOCKHOLM

	LW	HW	LW	HW
January 1, 1774		3.0 m		

It's not easy to be sure when you've found the seafront in Stockholm. The Swedish capital is built on an archipelago where a large lake, Mälaren, meets the Baltic Sea. On this bright February day, there is ice everywhere I look. The freshwater lakeshore is frozen solid; perhaps it is even cold enough for the brackish waters of the Baltic to freeze.

It is the vast cruise ships that finally help me get my bearings. They can only lie on the sea side of the various bridges linking the islands. The salt water is indeed trying to freeze. In shady corners, the sea ice makes chirruping noises where thin sheets scrape against one another. Larger pieces moved by the gentle swell from passing boats crack and squawk like jackdaws.

I am in the city to give a lecture, but I have a little free time left over and have startled my hosts by asking for directions to the city's long-standing tide gauge. Unsurprisingly, they are not sure where it is. I add that it may be close to the Slussen. This does provoke recognition, and some demur. It's as if I've said I want to see the London sewers or stroll through Central Park in the dark.

In Viking times, Mälaren was an extensive inlet that allowed seagoing vessels to venture far inland. But gradually the waterway became too shallow, owing, as we now know, to postglacial rebound as the land rose when relieved of the weight of ice upon it after the last ice age. Stockholm itself profited from these geophysical changes, finding itself positioned conveniently at a junction of the sea and a new freshwater lake. A lock was constructed in the seventeenth century, the Slussen, between the islands of Gamla Stan and Södermalm, to regulate the flow of water between the lake and the sea. Before long, however, the level of the lake had risen so far that age-old farmland became liable to flooding. Farmers' protests prompted a royal inquiry, which instigated the first scientific investigation of the water levels, as well as the construction of a new lock. At first, the sea level was noted

occasionally on a stone scale, but in 1774, systematic records began to be kept, with readings taken about once a week.

The tide gauge's successor today has been moved from its original location to a more salubrious spot on the nearby island of Skeppsholmen. It is a simple octagonal clapboard hut like a sentry box mounted out over the water. It looks quite at home among the tall ships moored along the same quay. Steps lead up to a door on one side. I try to peer in but cannot see the mechanism. At the site of the original sluice, meanwhile, modern metal lock gates lurk in the shadows under a busy overpass. It would make a promising crime scene in a Nordic noir drama.

Perhaps you think it counterintuitive that there should be a tide gauge, or mareograph as it is more formally named, in a place where there is so little tide. The whole of the Baltic Sea is almost tideless, owing to the constriction of the Danish straits. In fact, it makes perfect sense, not only for its purpose when it was installed—for the more efficient control of the height differential between Lake Mälaren and the sea—but perhaps even more so today, as we shall see.

Amsterdam (which then would have had a tidal range of about 1.5 meters) began regular measurements much earlier, in 1700, but the record was discontinued in 1925. Marks were scored into the side of the Doge's Palace in Venice in 1732, but not on a regular basis thereafter. The dockmaster at Liverpool (tidal range up to 8 meters) started a series of readings a few years before the Swedes, but the practice was only intermittently continued. Thus, it is Stockholm that possesses the longest continuous series of sea-level readings of any place in the world, despite having tides measured in single figures of centimeters. The figures may be of little practical use to mariners, but it is, in fact, precisely because the tides are in effect neutralized in these data that they are invaluable to scientists interested in an even more fundamental property of the interconnected global ocean: its height.

The equipment needed to measure the height of the tide is obviously the same as that needed to measure the level of the sea, since the tide is no more than a short-term cycle of variation in sea level. It just depends on how often you take your readings.

The earliest tide gauges were simply versions of the gradated posts stuck into the water that are still a frequent sight in coastal inlets and harbors today. Such a tide staff can be read as often as anybody cares to consult it. This was, in essence, the technology available to Galileo and Newton when they made their studies of the tides. As the science advanced, however, so did the demand for data, and it became important to have more or less continuous readings of the tide height over twenty-four-hour days for periods of weeks or even years. As international maritime trade expanded, collecting these data became an important, but very time-consuming, duty for many harbormasters around the world, contributing greatly to the scientific understanding of the tides, as we have seen. It was not until 1831 that one Henry Palmer, the civil engineer at the London Docks, devised a machine that would do the job automatically, using a float placed inside a small reservoir, known as a still well, connected to the sea, that maintained the same level while eliminating the movement of waves. A wire runs up from the float and over a wheel, which is geared in such a way as to convert the tidal rise and fall into movements of a pen that traces a record on a rotating paper drum. More recent devices use compressed air, radar, acoustic beams, or lasers to obtain more accurate readings, which are then converted into digital formats.

In addition, it is now possible to measure sea level remotely by satellite, using radio signals bounced off the sea surface. Analysis of these readings enables reliable estimates to be made of the average sea level anywhere in the ocean, whereas previously, data collection was limited to locations where deep-sea monitoring buoys were anchored. The readings also provide a profile of sea-level deviations from the average owing to tides and other effects. Recent satellite launches mean that both the quantity and quality of data available from altimetry have increased greatly even since the turn of the twenty-first century.

However, as often happens with new data, this innovation has made the picture more complicated. The tides may be thought of as a set of time-dependent cyclical variations away from average sea level produced, as we have seen, by an alarming number of gravitational effects due to the moon and the sun (as well as by a range of more minor local gravitational influences, as we shall see shortly). But it turns out that the idea of "sea level" itself is strictly relative. There is clearly an average sea level in any location where one cares to position a tide staff or a buoy packed with electronic instruments. But when these data points are compared globally, they do not tally. To put it simply, the sea is not, in fact, level at all. In some places, the average sea level is higher than in others all the time. This is nothing to do with the tides, but a permanent distortion that would never come to light by taking local measurements, which require the world-seeing eye of the satellite. Furthermore, the *global* average sea level, if we can agree on a reference datum, is not fixed but is itself changing.

We now have a mighty paradox: not only can we predict theoretical tides to millimeter accuracy—with far more precision, in fact, than we can ever say they will occur, because the figures are sure to be thrown off on the day by short-term weather effects that are not part of the tidal calculation—but we also now know that these tides ride on a misshapen ocean with irregularities in its height far larger than any tide. Is there, then, any point in this pursuit? Is there something else these data have to tell us?

SCIENTISTS HAVE SOPHISTICATED INSTRUMENTS and years of carefully maintained data to tell them what is happening to our seas. But where are the rest of us supposed to look for clues to these subtle shifts of the oceans—changes on timescales far longer than the tidal cycle and far smaller in magnitude, and yet possibly of far greater importance? Certainly, we cannot tell anything by staring at the horizon. It is surely one of the chief terrors of the sea that it lacks any solid point of reference. Are there signs where the sea meets the land? We have seen how large towns have been lost to the sea, while others have found their seagoing livelihood taken from them as the retreating sea leaves their harbors dry.

We have seen that paintings provide scant evidence of the workings of the tides. It is surprising, then, to find that they can provide useful information about longer-term changes of the sea and land. In the first decade of the twenty-first century, the Crown Estate, the UK agency that manages much of the country's seashore, intertidal zone, and seabed, launched a major survey of historic paintings and drawings of the coasts of Britain made since 1770, when the coast was coming to be seen as an amenity and therefore was a newly popular subject for artists.

Comparison of these impressions with the present situation can reveal important clues for geologists and oceanographers, as well as social scientists interested in our changing attitudes toward the coast. Artists may have favored the crumbling cliffs of East Anglia because they offered a portentous metaphor for social decay, but their work is now important because it offers documentary evidence in studies of coastal erosion. An especially appealing subject for a time was the Eccles church, which, having lost first its nave and transept to the waves, fetched up on the beach as the tower alone, still upright, a sentinel that stood for a generation until it, too, was finally consumed by the sea in 1895.

Occasionally, a single artwork tells a more elaborate story. In 1830, Charles Lyell published his *Principles of Geology*. His almost literally earth-shattering work argued that the best evidence of the nature of the earth's past was to be deduced from scientific study of its present geology. To this end, he traveled to Italy to study the volcanically active areas around Vesuvius and Etna. He came to the conclusion, unpalatable to some, that the planet must be hundreds of millions of years old, far older than the six thousand years indicated by the scriptures.

For the frontispiece of *Principles*, Lyell chose an engraving of a Roman ruin at Pozzuoli near Naples, then believed to be a temple to Serapis, though now thought to have been an indoor market. It shows the bases of three remaining columns awash in the sea. A third of the way up each of the columns is a sharp horizontal line. Above the line, the stone is darkened and coarsely etched, while below it is pale and smooth. Lyell identified the damage above the line as caused by a species of burrowing clam, while the lower parts must have been preserved by being buried beneath a layer of volcanic sediment. This could only mean that the building, initially on dry land of course, had first been partially buried during a volcanic eruption and had then sunk beneath the sea, only to rise again at a later date to reach its level when the artist drew it. The ruins and their revealing marks may still be seen today.

What this iconic image could not show, however, was whether it was the land that had moved down and then up, or the sea that had moved up and then down, or some combination of the two, although Lyell the geologist was adamant—and, as it turned out, correct—that the less believable possibility was true on this occasion: the land rose and fell in vast upheavals generated by violent forces deep in the earth.

IN 1834, Lyell toured Scandinavia, where the Baltic shore was to furnish him with another epiphany. More than a century before, the famous Swedish physicist Anders Celsius—known to most of us now for the scale of temperature named after him—had been drawn to investigate the apparent fall in the level of the sea in the Gulf of Bothnia, which locals noticed because seals no longer hauled themselves out to bask on previously favored rocks. Celsius selected four former seal rocks around the gulf, identified as such because they were mentioned in legal documents as commercial assets that had once provided an easy seal harvest. A similar rock survives on the island of Lövgrund off the coast of central Sweden, with the date Celsius scored into it, "1731," and a line underneath to mark the sea level. By comparing results from all four rocks, Celsius confirmed that the sea level was indeed falling throughout the Gulf of Bothnia. His findings excited other scientists, Linnaeus among them, but were resisted by clerics, who feared their implications for the biblical version of the earth's creation.

In his *Principles*, Lyell had assumed that dramatic vertical movements of the land could be produced only by volcanic action and earthquakes, as had happened at the Temple of Serapis. He initially refused to credit the idea of more gradual and less violent geological uplift—until, that is, he visited Celsius's rock on Lövgrund in a land that was, as he was forced to admit, "remarkably free within the times of history from violent earthquakes." He found that the sea level lay 87 centimeters below Celsius's mark made a century before. Today, the Celsius rock is scored with many lines and dates showing that the land continues to rise. When the time comes to inscribe the line for 2031, the rock will be almost on dry land.

But again, the question arises, What's really changing? Is this the sea level falling, as Celsius thought, or an uplift of the land, as Lyell overcame his skepticism to believe? And how can we tell for sure? In fact, there was no way to do this until the identification, later in the nineteenth century, of further marine sediments at various elevations above sea level, which enabled geologists to date the ice age and estimate the rate of rebound. Long before anybody dreamed of ice ages, Celsius had made the first measurements of postglacial rebound of the Fennoscandian landmass.

Today, the lines on Celsius's rock hint at a subtle secondary effect. The rate of fall in sea level, or rise of the land, is not quite linear; the trend seems to be very gradually slowing. The annual change observed by Celsius and Lyell of more than 8 millimeters is now down to less than 7 millimeters. Is this massive geological correction suddenly coming to a halt? Or is something else afoot? The Baltic has news for us yet.

ON STRANGER TIDES

"Go where the data is," said one geophysicist I spoke to. "That's a really good piece of advice." For a scientist, to be sure. But I am conscious that my topic is already complicated enough, and I should perhaps not chase down every last ingredient of the tide. Nevertheless, today it is these additional wrinkles—effects that may produce sea changes measured in centimeters or millimeters only—that are driving renewed interest in research into the tides, and exciting scientists from many disciplines concerned with the interconnected workings of our planet. In the nineteenth century, it was a surge of measurements from ports around the world, tediously made by dutiful harbormasters, that gave physicists and mathematicians what they needed in order to refine their theories. Now, it is high-resolution data streaming from many sources, both close and remote, as well as detailed computer modeling, that promise to bring our understanding of oceans to a new level of sophistication.

In the broadest scientific definition of the word, "tide" refers to the motion of a fluid under the influence of the force of gravitation. The sea is only the most familiar fluid. The air and even the earth itself are also manipulated by the astronomical pull of the moon and the sun. Newton was the first to conjecture that there were tides in the atmosphere as well as the oceans, and Laplace hoped that the atmosphere, if not the seas, could be shown to behave in conformance with his theory of tides. But it was the Prussian explorer Alexander von Humboldt who was one of the first to actually detect the "regularity of the ebb and flow of the aerial ocean undisturbed by storms, hurricanes, rain, and earthquakes," on his voyage to the American tropics in 1799. In these latitudes, the stable weather meant that atmospheric pressure usually varied little, enabling a small, twice-daily oscillation in pressure of only 1 millibar or so to be detected.

Soon afterward, scientists such as Lord Kelvin began to reason that the earth itself must also yield in response to gravitational forces, although he was skeptical of the new idea that the earth's crust might be formed around a molten core, which clearly supported this theory. The existence of earth tides was proved by George Darwin somewhat later, in 1882. Naturally, these massive movements of the atmosphere above and the earth below in turn introduce yet more complicating factors that must be taken into account in a full calculation of the ocean tides.

In addition to these new tides, there are sea tides of a kind that we have not yet considered. At first, these were just "noise," gremlins inexplicably distorting measurements of properties such as ocean temperature that scientists had supposed to follow a smoothly predictable pattern. Eventually, the anomalies were found to be an effect of the movement of boundaries between various deep layers of the ocean, which become stratified because of differences in temperature or salinity. (The tides we are familiar with may be thought of, likewise, as a movement of the boundary between the fluid layers of the sea and the air.) These "internal tides" were first identified by the famous Norwegian Arctic explorer Fridtjof Nansen, who used bottles lowered on wires from the side of his ship to measure the water temperature and salinity at different depths. They do not move in phase with the surface tides, but have their own speed and scale, like the oil and vinegar slopping back and forth at different rates in a bottle of salad dressing. Rob Hall, a continental-shelf oceanographer at the University of East Anglia who has become interested in these deep-ocean tides, prefers another analogy. "Whenever I'm in a pub talking about this, I try to get a pint of Guinness, because you can make waves between the beer and the foam," he explains.

The vertical range of internal tides is much greater than that of surface tides, typically reaching up to 30 meters. The largest tidal ranges have been measured in the Luzon Strait between Taiwan and the Philippines, where the difference between internal highs and lows can be as much as 700 meters in a 4,000-meter depth of ocean. Occasionally, an internal-tide wave will be so large that it reaches the ocean surface. When I first heard this, I had high hopes that this surprising phenomenon might explain the disappearance of ships in the Bermuda Triangle or some other great mystery of the sea. But no such drama. Rob tells me that the only visible evidence of this deep ebb and flood might be the occasional sight of bands of smooth water lying across the calm ocean.

The news of internal tides has awakened the interest of oceanographers, but also of climatologists and marine biologists, because they may be involved in the transport of marine life, sediments, and minerals, as well as of gases and heat from the atmosphere. The quiet kinetic energy of internal tides is also stolen from the more obvious surface tides. The undersea currents produced by internal tides may provide a horizontal vector of movement for these materials at speeds and in directions that had not been imagined, while the vertical movement of water must affect the exchange of heat between the atmosphere and the ocean, with implications for our understanding of the likely impacts of anthropogenic climate change.

All of these tides—of the air, of the earth itself, and of the layered depths of the sea—must be taken into account in full calculations of the familiar tide that alters the level of the sea. And they, too, must be understood and measured if we are to arrive at a meaningful idea of the true sea level from which the tides are merely a deviation.

ISOLATED VARIABLE

The range of the tides at Stockholm is so low—about 3 centimeters—that the expected twice-daily pattern we expect to see in most tidal waters is frequently subsumed by other effects. Some tide tables for the city are published, but they often show an entire day or two passing without a recorded high tide at all. Surely such minute and irregular movements can be of only theoretical interest. We are now finding, however, that it is good these data are kept, not because they offer any useful information for sailors, but because of the long-term trends they unwittingly record.

In 1891, after the Stockholm authorities had been monitoring the sea level for more than a century, the engineer in charge of the city's harbor and bridges, Victor Lilienberg, analyzed the accumulated data. He found that the sea level was progressively falling, just as Lyell had found at the Celsius rock. But he also observed that the *rate* of decrease seemed to be gradually leveling off too. This could mean two things: either postglacial rebound was suddenly slowing, or there was a new effect from a rising sea. Lilienberg had perhaps stumbled on the first evidence of rising sea levels owing to changing climate, because his data happened to span a period of transition from the end of the Little Ice Age into the beginning of a warmer period, which was accompanied by the melting of glaciers.

Because the sea-level rise is so small compared to the postglacial uplift of the land, it would be easy to dismiss it as insignificant. However, this is where long data series really come into their own, allowing statistical significance to be established. The run of data at Stockholm—well over two hundred years now—shows that this small rise superimposed on the constant fall is not a transient effect or an error, but a real and continuous natural phenomenon. This trend reflects a warming of the climate that now includes the contribution owing to the emission of greenhouse gases.

For the stories of Celsius's and Lilienberg's studies of Baltic sea levels, I am indebted to the Swedish geophysicist Martin Ekman. Somewhat unusually for a modern scientist, Martin is the director of his own independent research foundation, the Summer Institute for Historical Geophysics, which is based in the officially stateless Åland Islands in the Gulf of Bothnia equidistant between Sweden and Finland.

The islands have an interesting history. At the beginning of the nineteenth century, they were claimed from Sweden by Russia, seeking a harbor for its Baltic fleet that would be less susceptible to flooding than its major naval base on Kronstadt Island, where the sea can rise to dangerous levels whenever sustained westerly winds blow and large volumes of water are forced into the Gulf of Finland and the tight confines of the Neva Bay that form the seaway to St. Petersburg and the naval base. The islands were demilitarized after the British and French attacked them during the Crimean War. A fortress built by the Russians at Bomarsund on the main island was destroyed during the battle, although a stone tide gauge there still survives; like a number of others dotted around the Baltic, the tide gauge now finds itself stranded some way above the sea, however. Åland was then administered as part of Russified Finland. After the Finnish Civil War in 1918, the islanders hoped to secede and integrate with Sweden, but the Finnish government refused, and the dispute was referred to the League of Nations. Today, the islands are the sovereign territory of Finland, but they are Swedish-speaking, autonomous in government, and have their own flag and Internet country code, .ax; they remain demilitarized.

I imagine Martin Ekman's attachment to the scientific establishment as being similarly ambiguous. His institute, it turns out, is really "myself and my small house." When he's not working there, he is sometimes to be found perched on a rock by the sea. "I started it because I like to do what I want, freely working with science and history," he tells me. "I'm not the kind of person who wants to spend a lot of time dealing with bureaucracy. I want to spend my life writing and thinking." The papers he issues from time to time are occasionally digressive—one, for example, including astronomical calculations to establish whether the Vikings who reached America could have seen the midnight sun from their landing place as related in the sagas (they could). But the bulk of his research is focused entirely on one variable—the sea level in the Baltic. "From this quantity, as observed during three centuries, we are able to draw conclusions about the behavior of our earth as a whole: its interior, its oceans, and its atmosphere," he explains assuredly.

How can one man, in isolation, draw such bold inferences? This is surely the dictum "think global, act local" taken to its extreme of absurdity. But Martin is adamant: "This single physical quantity—the Baltic sea level—tells us a lot about the planet we live on. You can study it with your own eyes—and it is beautiful—yet it is so complex too." Although his institute may seem remote, I find he is, in a way, at the center of things. Baltic shores are well supplied with reliable records of sea level, both from the tide gauges of the scientific era, and from a time before those became available, from the evidence of Viking and earlier sites that once stood on the coast and now find themselves landlocked. If we can decode their message, these historical records stand to inform our scientific understanding not only of postglacial rebound, but also now of rising sea levels related to climate change. They may also reveal certain nuances of the tides that cannot easily be monitored in more lively seas, and even the topography of the sea surface itself, which, in addition to everything else, has subtle deformations that arise chiefly from differences in salinity from one place to another, differences that are usefully exaggerated in the Baltic Sea.

THE NUB OF ALL THESE PROBLEMS is still knowing what sea level actually is. The sea is moved vertically by the tides, as well as by various other effects. Underlying this motion there may also be vertical movement of the earth's crust, which must be taken into account if we wish to know the true action of the sea. Add to this the fact that the sea is not, in fact, level in itself, even in the impossible event of an utterly calm and tideless day. It has short-range waves raised by winds and the longer ocean swell. It has the sea-wide slop that troubled the Kronstadt fleet. It has the tides in all their complexity, which, as we have seen already, may be high in one place at the same time as they are low in another, and are never at rest. Yet it still requires a datum of sea level to which all concerned can agree before we can talk meaningfully about the height of these waves or tides, or for that matter the height of any point on land, which we casually speak of as being so far "above sea level."

It is a relatively simple matter to establish a datum for one place. The UK datum is based on the average or mean sea level at Newlyn in Cornwall, for example—or, to be more exact, the mean sea level at Newlyn as measured between the years 1915 and 1921. In contrast to locations in the Baltic, the sea level here has risen by about 20 centimeters since that datum was chosen, but it is still this historic figure that is used. This datum assumes that there is no significant vertical movement of the earth's crust to factor in, which is a reliable enough assumption in many places not subject to postglacial rebound. Other countries naturally choose datums located in their own territory, including Stockholm, Helsinki, Copenhagen, Kronstadt, and the Prussian port of Swinemünde (now Świnoujście in Poland), used by their respective Baltic nations. In some of these locations, however, the land is rising quite significantly, and so the standard datum is constantly changing relative to places where the land is stable. This means that net changes in sea level must be measured according to this local sliding scale. National interests have been overcome in North America, on the other hand, where the datum is, in effect, an average of averages involving measurements of the sea level made at ports in Canada, Mexico, and the United States. European countries, too, are now turning away from their old nationalistic approach, and beginning to use Amsterdam as a common datum.

To me, these arrangements all sound rather ad hoc and unsatisfactory, and a long way from providing the objectivity that science requires. Fortunately, there is an alternative to these traditional measures. We can instead define sea level in the ideal way that a mathematician or a physicist would choose, as the surface through which the direction of gravity runs everywhere exactly perpendicularly. A few decades ago, this would have been a pointless nicety. But it is now possible, using the combined forces of satellite altimetry and tide gauge measurements, to adopt this ideal surface—scientists call it the geoid—as a global datum. The geoid never actually appears. It is the shape the oceans would settle to if the earth stopped moving, if there were no ocean currents, no tides, and no weather. It is not a model of perfection, however. It does take into account permanent geophysical irregularities, such as the fact that the solid earth is not a sphere but is roughly ellipsoidal and its interior is not uniformly dense. This uneven distribution of the earth's mass alters the planet's own gravitational field, which in turn affects the base level of the sea. In fact, the global ocean bulges by about 10 meters at the North Pole and has a dimple of about 30 meters at the South Pole. Elsewhere, there are even larger irregularities, owing to underlying asymmetries in the earth's core. Even features such as undersea mountain ranges introduce their own smaller distortions of the sea level directly above them. It is this decidedly non-ideal, dented and lumpy ocean surface that today constitutes the ultimate baseline for measurements of changing sea levels over all periods of time, from the millennial shifts of the ice ages to the hour-by-hour ebb and flow of the tide.

It is clearly no longer adequate to rely on a mean sea level prescribed by one nation or another. Instead, this key datum is carefully tended by an international body, the Permanent Service for Mean Sea Level (PSMSL), based at the National Oceanography Centre in Liverpool, which processes measurements from two thousand tide gauges based on the coasts of all the continents and on various mid-ocean islands. The global approach has revealed new anomalies, such as height differences between parts of the Baltic Sea and the open oceans, owing to the low concentration of salt in the former, which makes the water locally less dense. In general, the brackish waters of the Baltic are up to 30 centimeters higher than the saltier waters on the North Sea side of the Danish straits. There is clearly an element of arbitrariness in all this. Nevertheless, the globally agreed-upon figure established by the PSMSL provides an essential and perfectly adequate benchmark against which changes in sea level and new tidal effects may be measured.

WHAT CAN MARTIN EKMAN contribute to this urgent global matter from his island redoubt? As he explains to me, his science gains as much from history as it does from geography. It is not only the fact that tidal variation is, for all intents and purposes, eliminated here, meaning that other changes of sea level can be studied more easily. It is also that these coasts have some of the longest series of sea-level records available anywhere, series that are densely repeated in many locations up and down the Baltic, the keeping of which since the eighteenth century has been prompted not by an interest in the tides, but by an interest in the postglacial uplift of the land that is such a characteristic of the region. "This combination seems to have created the special situation where Nordic people have contributed considerably to discovering all sorts of nontidal sea-level changes," says Martin.

In making predictions about global climate change, scientists are acutely aware that, for their claims to be credible, their data must be beyond reproach. Yet we have only the past to help us predict the future. The comparative wealth of historical data available from these shores is a vital asset for climatologists everywhere, enabling forecasts that may influence the way we all live in future to be made with greater confidence. Martin's accidental discovery of Stockholm's historical tide gauge records thirty years ago—he found them irregularly filed in the city archives—came at a time when the Stockholm tidal station was threatened with discontinuation; he may have saved the future as well as the past.

As for me, I am reassured to have found a scientist who can work in one place and, through it, feel a profound connection with the workings of the planet as a whole.

SMALL CHANGE

Making use of figures supplied by the PSMSL, the United Nations Intergovernmental Panel on Climate Change concluded in its 2014 assessment report, "It is very likely that the mean rate of global averaged sea level rise was 1.7 [1.5 to 1.9] mm/yr between 1901 and 2010, 2.0 [1.7 to 2.3] mm/yr between 1971 and 2010 and 3.2 [2.8 to 3.6] mm/yr between 1993 and 2010. Tide-gauge and satellite altimeter data are consistent regarding the higher rate of the latter period." (The figures in brackets represent confidence limits.) Overall, the IPCC scientists predicted that the average sea level will rise by at least 28 centimeters by the end of the century, and could rise by nearly a meter. These rises may not sound very big, but they must be set against the fact that more than a billion people live in low-lying coastal areas, with as many as a hundred million currently living below 1 meter above sea level. A global increase in temperature of 2°C during the brief interglacial period 120,000 years ago is thought to have caused sea levels then to rise by between 5 and 10 meters as much of the Greenland and Antarctic ice sheets melted into the oceans. Two degrees is the increasingly hopeless-looking target maximum average global temperature increase set during the 2009 United Nations summit conference on climate change in Copenhagen.

The present rate of increase is actually rather less than the long-term average seen since the end of the last ice age, about twenty-five thousand years ago. Since then, the oceans are thought to have risen by more than 130 meters, an annual average rise of 5 millimeters. The melting that produced all this extra water was largely complete six thousand years ago, though, and the natural rate of sea-level rise has tailed off to less than 1 millimeter per year. Since the early part of the twentieth century, however, this natural rise has been augmented by a new contribution, which most scientists attribute to warming produced by the emission of greenhouse gases, and which now appears to be rising exponentially.

Skeptics about climate change have challenged the scientific data on rising sea levels. They contest the rate of the rise and have been especially critical of this recently claimed increase in the rate. The IPCC itself has admitted that it is hard to be sure of the benchmarks, and is careful to select tide gauges based in stable land regions. There are more fundamental uncertainties too. Large portions of the sea-level rise observed since 1900 can be safely attributed to melting glaciers and thermal expansion of the oceans, but about a third of the rise is unaccounted for. The fact that the IPCC, in its efforts to use the best data available from all sources, has deduced the trend by combining data from tide gauge measurements and more recent satellite altimetry has also excited skeptics' suspicions of the scientists' methods.

However, the skeptics make basic errors that show up their ignorance of the subtleties of scientific data. The claim that "Stockholm [sea level] is actually falling," triumphantly repeated on many Internet forums, lacks any understanding of—or simply chooses to deny—the contribution to this figure made by the postglacial rebound. Martin Ekman is astonished when I tell him of this misrepresentation. "Within the Nordic area nobody would be able to state that," he tells me. "Even children know about the uplift phenomenon."

The predicted figure for a global rise in sea levels is an average that takes into account extremes, including those places in the Baltic where the local sea level is actually falling, as well as other places where it is rising faster than average. By 2050, the British government expects to see a sea-level rise of just 17 centimeters in Scotland, but 37 centimeters in East Anglia, for example. In parts of Alaska, the sea level is falling even faster than it is in the Baltic, while in Louisiana it is rising very fast, by a centimeter a year—three times the global average.

Additional physical effects that influence local sea-level changes can be hard to predict and sometimes seem highly counterintuitive. In general, for example, sea levels may not rise most near melting ice sheets, where you might expect them to; there, they may fall because of the lower gravitational attraction exerted by the reduced ice mass (another "tide" to add to the list). Correspondingly, it is far from the poles that the sea level is expected to rise disproportionately fast. A paper published in the journal *Science* by geophysicists at the University of Toronto in 2009 found that sea-level rise would be greater than previously estimated along US coasts, and cheekily suggested that, if the West Antarctic Ice Sheet were to melt, Washington, DC, would be among the places most endangered.

The ocean is increasingly recognized as an important factor in global climate change. It exchanges gases and heat with the atmosphere in ways that are not yet fully understood. The tides make their own generous contribution to this uncertainty. Their rise and fall clearly involves the interconversion of massive amounts of kinetic and potential energy. But one huge question remains: where does the energy go in the end? Burntcoat Head and the Saltstraumen gave me a glimpse of the colossal power of the tide. Its energy, captured from the orbits of the moon and the earth, has to be dissipated somewhere. The whirlpools and turbulent currents I saw hinted at this process in action, as the kinetic energy of the water was converted into heat by the friction of its waves and currents against one another, and against the air and the land.

But this is not the half of it. Somehow, the tides must dissipate all the power they gain, which is calculated at about 3 terawatts (3,000,000 megawatts). Estimates for the power that the tides lose when they interact with the coasts of all the continents come to a mere 20,000 megawatts, less than 1 percent of this total. The friction between layers of seawater forced against one another by internal tides may account for some more losses, but it is thought that the bulk of the energy must be exchanged along midocean ridges. Knowing the answer to this unresolved question may influence the planning of schemes to extract tidal power, although, while we are bandying large numbers around, it is worth recalling that the global tidal-power dissipation, huge though it is, is less than a tenth of that available from geothermal sources, and a tiny fraction of the solar power constantly striking the earth.

In due course, changing sea levels may produce alterations in familiar tidal patterns. Already, oceanographers have begun to observe that tides are evolving in new ways that cannot be explained by astronomical gravitational forces on their own. Of course, sea-level rises will everywhere be small compared with the typical range of the tides—a truism that had led some scientists to consider them an unimportant influence on tides themselves. However, computer modeling now suggests that even small changes in the overall depth of the ocean might lead to significant tidal changes in some locations. Naturally, the picture is not so simple as greater tides occurring where the sea level has risen most; scientists have, in fact, found no simple correlation between the extent of sea-level rise and change of tidal pattern from place to place. Thus, where the tides are dominated by a local oscillation—in a large, shallow bay, for example—a small change of depth can produce a relatively large change in the total volume of water flowing back and forth. This volume might be great enough to alter the natural frequency and the amplitude of the oscillation, thereby affecting both the timing and the height of the tides.

In some areas, these effects may be large enough that they ought to be taken into consideration for navigation and in the planning of sea defenses. For example, if a projected sea-level rise of half a meter happens to reduce the height of the highest tides by half a meter, then there may be no need to improve coastal protection measures, as might have been anticipated. On the other hand, if it happens that a locally *falling* sea level will bring about a *greater* tidal range, then there may unexpectedly be a case for action in a place that was thought not to be threatened. Over recent decades, the tides in parts of the North Sea have been increasing at a rate comparable with the increase in sea levels, for reasons that scientists do not yet fully understand. Thus, in the river port of Hamburg, for example, the range of the tides is expected to increase, exacerbating increases already observed during the past century attributed to dredging. Meanwhile, Dublin, which experiences substantial tides today, may one day cease to be a tidal port altogether as rising sea levels cause an amphidrome (where there is no vertical tidal movement) presently 45 miles to the south to migrate toward the city.

Such changing tidal patterns may also affect the long-term prospects of tidal-power installations. For example, there is presently a plan to enclose part of Swansea Bay in Wales and make it into a lagoon in order to extract power using the large difference in height of the seawater at different phases of the tidal cycle. (Schemes to harness the huge tides in the Severn estuary have a history stretching back nearly a century.) The 6-mile barrage would have a generating capacity of 320 megawatts, enough for 155,000 homes, satisfying the demand for domestic electricity for the city of Swansea and beyond. Unlike more radical schemes to extract power by positioning undersea turbines in fast-moving tidal streams in places such as the Bay of Fundy and Orkney, the technology is proven. Indeed, a somewhat similar plant has been in operation for more than fifty years at La Rance in Brittany, where the tides are almost as great. However, if the tidal range is reduced as a consequence of sea-level rise during the 120-year lifetime of the barrage—and one recent study predicts a 10 percent reduction, in the event of a 2-meter rise in sea level—then the installation may produce rather less electricity than first thought.

THE SEA-LEVEL COMMITMENT

The greatest impact of rising sea levels and the changing tides that may accompany them will be on human habitation. The evidence over tens of thousands of years suggests it is unlikely we will relinquish our place by the sea just because the level is changing. The numbers of people living close to the sea are rising, and so—even faster—are the numbers exposed to the risk of coastal flooding. Accurate knowledge both of future sea levels and of future tides is essential if the coastal defenses needed to protect these communities are to be cost-efficient.

Ironically, it is our love of a sea view that exacerbates the problem in many wealthy parts of the world, where the priority has become to protect individual properties right down to the water's edge rather than to maintain a resilient seashore for all. The natural tendency on coasts where the sea level is rising is for beaches to rebuild themselves further inland as the sea forces them back. The erection of solid defenses such as concrete seawalls in front of seaside properties prevents the sand from shifting inland, and may well hasten the disappearance of the very beach that made it attractive to build there in the first place. It is likely that some expensive ocean-facing homes in places such as Australia's Gold Coast, Dubai, and Long Island, New York, are at risk in this way. Our perception of this geophysical process as a threat rather than a simple change is unintentionally revealing about our idea of property, which culture and law decree must be permanently fixed, even though the area of land with which it is legally identified may itself be on the move. We have yet to learn the obvious lesson that for a property to remain beachside, it must go where the beach goes.

In general, we are fortunate that the life cycle of most buildings permits a measured response to rising sea levels. Some established coastal settlements may be gradually rebuilt to make them more resilient—for example, by putting houses on stilts or tethered floating platforms, or by zoning the land use in such a way that an occasional tidal surge is a manageable inconvenience lasting a few hours, and not a life-changing catastrophe. In other locations, the sensible course of action will be "managed retreat," a phrase that has been criticized as a politician's euphemism for simply giving up, but that fairly reflects the fact that it would be folly to build new homes in areas liable to suffer from coastal flooding in future.

Even if we soon stop burning fossil fuels and achieve a new equilibrium of carbon dioxide in the atmosphere—highly unlikely at present trends—global temperatures are certain to keep rising for some time, and sea levels are certain to keep rising for even longer, as the heating of the atmosphere is spread gradually through the oceans and accelerates the melting of the polar ice sheets. This sea-level rise—which scientists have termed the "sea-level commitment"—is, in effect, already locked in for the future. A 2°C increase in global temperatures in the short term is likely to produce a sea-level rise of up to 5 meters, although the full effects may not be observed until hundreds of years hence. And with this overall change in sea level will come less predictable changes in the extreme levels experienced during storm surges, which are already doing damage to our coasts. In the long term, if not the short, "managed retreat" is our only option. The sea always wins in the end.

13

DILUVION

"Shingle Street is Special. Only you can keep it that way," says a sign placed by the Suffolk County Council and "The Landowners" at the top of the beach. It is indeed special—a designated Site of Special Scientific Interest, in fact—and a rare and bleak habitat where the graded, rubbed-round pebbles that give the place its name stretch fully 300 yards from a lonely row of coastguard cottages down to the edge of the sea.

Across the pointillist gray and orange and black of the shingle runs a line of bleached whelk shells, one whelk wide, all the way from the water to where the marram grass finally gains a purchase and the land can be said to begin. I have seen it before, and wondered then who made it and how long it has been here. What is its story? Is it a memorial of some kind? Does it trace some ancient path?

It is interesting to me now because it helps to illustrate the scientific curiosity of this beach. The shell line picks out the fact that the shingle is not a featureless expanse, but is arranged in seemingly endless stony waves. The white line recedes in side-slipping bursts, like a road seen disappearing across a rolling prairie. The waves vary in every quality: in their height, in their distance from crest to crest, in their slope up and slope down, and in the size of pebbles on each. They may be read a little like tree rings or the compressed layers of

snow in ice cores. Each ridge is evidence of a major storm. The spacing between the ridges represents intervals in time, with those ridges close to the waterline, typically small but sharply peaked, formed by storms of recent weeks or months, and the larger undulations further back made by storms years and even centuries ago back to prehistoric times. Interpretation of these shifting geological forms is far from straightforward, but the height and steepness of each shingle bank also say something about the severity of the weather on each extreme occasion.

Just off the shore, more shingle is heaped in short-lived banks. Caressed by the run of the tide up and down the shore, it forms into little islands, as well as looping and hammer-headed spits that jut out into the sea as boldly as Victorian piers. In the 1950s, scientists at Orford Ness, the much longer and somewhat more permanent shingle spit to the north, where the Atomic Weapons Research Establishment once had a station, studied the pattern of movement of the shingle driven by waves and tides by coating individual pebbles with radioactive barium paint and then scanning the seabed and beaches for radiation hot spots.

It is not surprising that the place seems haunted. It is history in geography: the past arranged not, as usual, in buried vertical strata, but laid out in the horizontal sequence of shingle banks. Somewhere under these millions of pebbles may lie the dead of the secret Nazi invasion reputedly thwarted here in the summer of 1940, as well as the bodies of four coastguardmen never recovered after their boat got into difficulties following a river dash up to Aldeburgh for supplies a century ago. Perhaps the long-lost wreck of Captain Hewett's HMS *Fairy* is somewhere here too.

The shell line, I learn later, is maintained by two artists, childhood friends, one from Cambridge, the other from the Netherlands. They started it in 2005 as a recuperation project after serious illness and come back periodically to make repairs. But it is clear that most people who pass this way do their bit to ensure that the line is never broken.

SHINGLE STREET is one of those places on the Suffolk coast that I feel I have come upon as if remembered from long ago as a consequence of reading W. G. Sebald's *Rings of Saturn*. My present excursion will also take me in his footsteps to desolate Lowestoft and the forbidden cliffs of Covehithe. Sebald's wanderings were suffused with a characteristic melancholy. But when I retrace his steps, I sometimes find almost the opposite, a kind of reckless optimism. Even in Dunwich—already halved in size by the time of the Domesday Book in 1086—the museum informs me that the "last" church was lost to the sea in 1920, as if it's all over now and the remaining fragment of the town will not disappear beneath the waves in the next century or so. The rate of erosion here is constantly monitored and is presently decelerating, I'm told, owing to a high buildup of protecting shingle bank. Will it last? It would take only one big storm to sweep the shingle away, and then the sea would start to feed on the low sandy cliffs behind. Yet there is much noisy house building going on in the town.

The rings of Saturn were first observed by Galileo. Sebald notes in an epigraph the modern understanding that the rings are composed of ice particles that may be the pulverized remains of a glacial moon, whose orbit decayed until it was torn apart by the tidal forces of the planet beneath. It is a reminder that our tides on earth behave in the bounded way that they do because our own moon circles in a stable orbit.

As I cross the county border back into Norfolk, I stop in a manicured trailer park on the cliffs at Hopton-on-Sea. A large chunk of the cliff has been lost to the sea following a sustained battering by easterly winds and large waves in the early spring of 2013. A length of white plastic fencing meant to look like timber has disappeared with it, and a few of the hardstands nearest the edge have had to be cleared and their trailers moved to safer sites inland. It is early in the season yet, but I find one trailer owner to talk to. Tony confirms that the winter storms did the damage but adds his belief—shared with many on this stretch of coast—that the erosion is being exacerbated by alteration of the tidal currents caused by works to enlarge the outer harbor at the port of Great Yarmouth a little way up the coast. "This is what we get now—bang, bang, bang," adds his wife cheerfully, pointing out the diggers at the foot of the cliff grappling with boulders of granite on what remains of the beach.

The hope is that the massive rocks will break up future storm waves and protect the cliffs behind. But the rocks themselves may have unintended consequences. By preventing erosion of soft sand and cliffs here, they reduce the quantity of sediment that is naturally washed away and made available to be deposited somewhere else. I am put in mind of John Donne's famous *Meditation XVII*, the one that begins, "No man is an iland." The poet continues, "If a clod bee washed away by the Sea, Europe is the lesse." Here, though, parts of Europe are being transported from one country to another, and we might think it is no worse than a zero-sum game. But after all the heavy lifting, some place in Europe is still the less for it. And even locally, the net effect of the work is often simply to shift the focus of the erosion a few miles along the coast.

The work at Hopton-on-Sea has been funded by government agencies, as well as by the owner of the trailer park. But in some places along this exposed coast, the national government has announced that there will be no funds to bolster sea defenses. At the same time, some planning restrictions have been removed, and now additional protection measures are being funded on the promise of profits from coastal property development in places where it might not have been permitted before. It is a rather different interpretation of the message "Only you can keep it that way."

The obscure term for once solid land now judged to be permanently lost to the foreshore below the high-tide line is "diluvion." It is significant in law because it signifies the transfer of land from a onetime landowner to the owner of the foreshore, land that formerly had value for building or grazing but whose value as property has essentially been cut to nothing by the action of natural forces. In the United Kingdom, such land very often falls into the hands of the Crown Estate, which owns and enjoys the right to exploit much of the British foreshore.

"Diluvion" seems appropriately apocalyptic too, with its echo of biblical flood and oblivion. The UK Environment Agency predicts that seven thousand English and Welsh properties could be lost to the sea over the next century, eight hundred of them in the next twenty years. The government is resisting calls to compensate home owners because to do so might be taken as a precedent for all sorts of financial loss sustained as a result of climate change.

Tony tells me that Hopton-on-Sea's granite is brought on barges from Norway. Indeed, I recall seeing very similar boulders, though naturally formed rather than blasted out by dynamite, that had tumbled onto the beaches in the Lofoten Islands, broken away from the precipitous mountains by Arctic frosts. I am struck by the folly of eating away at the hard west-facing coast of Scandinavia and dragging its rock across the sea to protect England's soft eastern rump. This puny human redistribution of the earth's geology seemingly aspires to counter the very rotation of the earth, which is what ultimately generates the forces of erosion that are so unevenly apportioned along different coasts. But to me, it offers an unintended expression of the futility of trying to stop the sea. It is a futility that Sisyphus would understand all too well.

SHEER LUNACY

Musing on humanity's futile attempts to alter the cosmic order of things, I recall a military exercise that I may have heard about first when interviewing colleagues of Carl Sagan as part of my research for an unrealized book. At the height of the Cold War, a research foundation in Illinois undertook work for the US Air Force to explore the possibility of detonating a nuclear bomb on the moon. The work was code-named Project A119. The future popular astronomer Sagan was closely involved and, in 1958, dutifully produced a scientific paper with the title "Possible Contribution of Lunar Nuclear Weapons Detonations to the Solution of Some Problems in Planetary Astronomy."

The military's ostensible aim was to find an acceptable way of studying the explosive and radiological effects of bombs too large to detonate safely on earth. But it is easy enough to imagine genuine scientific problems that might be investigated by these violent means, from probing the lunar geology to observing the effect on the earth of nudging the moon in its orbit. When details of the top secret project were finally released in 2012, the newspapers wrote incredulously about the long-abandoned plan "to nuke the moon" and even "to blow up the moon."

Project A119 raises the "thought experiment" of what effect blowing up the moon or suddenly removing it from its orbit would have on the earth. It turns out that thinking about the moon in such radical ways is far from new. In Islam, for example, the moon is said to have split in two as the end of the world approached, in what is variously interpreted as a miracle performed by the Prophet or as a prophecy by him.

In 1991, an Armenian American mathematician named Alexander Abian at Iowa State University took the project further—theoretically speaking—by suggesting that the earth would be better off without the moon altogether. Not only would there be no tides, but the absence of a nearby gravitational influence would allow the earth's axis of spin to realign itself, eliminating the seasons as well. How this removal job was to be accomplished was never clear. Simply blowing up the moon, even if it could be done, would create large fragments, some of which would be bound to crash into the earth, obliterating all life. Abian, though, was unabashed, likening himself to the persecuted Galileo when fellow academics disparaged his theory.

The moon will answer the matter in its own good time, because it is not quite the constant feature in the sky that it appears to be. Although it is not known exactly how the earth's only natural satellite came into existence, it is certain that its initial orbit was once far closer than it is today. At first, the hot, new-formed earth and its moon (probably captured as the result of a glancing impact of another planet-size body) held one another in a tight embrace. The earth also spun faster on its axis, making the day perhaps only six hours long when the earth-moon system came into being, more than four billion years ago. Gradually, the two bodies' motions slowed, and they moved further apart. By about one billion years ago, the water vapor in the earth's atmosphere had condensed into oceans, and the early continents began to form. With the distance between the earth and the moon perhaps about three-quarters of what it is now, the large gravitational attraction between the two raised vast tides. The energy dissipated by these tides as friction continued the deceleration to the rotational and orbital speeds that define our day and month now, and the moon became locked so that it always presents the same face to the earth. Eventually, the earth, too, will become tidally locked in this way, with the same face always looking back toward the moon— although this might not occur for another fifty billion years or so, by which time it is thought the planets will in any case have been swallowed up by their dying star, the sun.

Scientists have measured the rate of lunar recession by means of laser light reflected from mirrors placed on the surface of the moon by astronauts. The distance is increasing by more than 3 centimeters a year. In a million years' time, the moon will have drifted into an orbit 30 kilometers farther from the earth. The rate of recession is greater than it has been at some times in the past because of the present disposition of the continents. They now lie largely in north–south bands, which run at right angles to the earth's direction of spin. This orientation serves to increase tidal friction, thereby more effectively slowing the earth's rotation, whereas the earlier supercontinents, such as Pangaea (the principal landmass three hundred million years ago that was concentrated in the Southern Hemisphere), presented far less of their coastal shelves to the onslaught of the tide.

THE RETURN OF THE SEA

BLAKENEY, NORFOLK

	LW	HW	LW	HW
December 5, 2013		08:04 3.6 m		20:19 3.7 m

It was two days after new moon on Thursday, December 5, 2013, when the East Anglian coast experienced its worst storm surge since the great flood of 1953. As on that earlier occasion, strong onshore winds coincided with the spring tide to raise the level of the sea well above what had been predicted. This time, however, the coasts that suffered most were those of Lincolnshire and Norfolk rather than those of Suffolk and Essex.

Globally rising sea levels—the rate of rise in this region being around 3 millimeters a year—were perhaps thought by some to be ignorably tiny. During the ordinary run of events, they were. But the fact now was that the sea was 18 centimeters higher to begin with than it had been in 1953—the equivalent of two brick courses on any seawall. Had the defenses been improved to keep pace with the growing risk?

In places east of Wells-next-the-Sea, midway along the north Norfolk coast, the surge was less severe than in 1953, but to the west, in towns such as Boston and King's Lynn, the waters rose higher. In Wells itself, new works are under way to raise the seawall around the harbor. On a brick column of the harbormaster's office—which lies on the seaward side of this wall—three stainless steel plaques record the floods of 1978, 1953, and 2013. The sea level in 1978 would have come up to my hips. The 1953 line is level with my ribs. The 2013 line is at my neck. Occasionally along the Norfolk coastal path, I have seen more informal marks put up by local landowners that seem tinged with anger and frustration at the inching progress of the sea. At Wells, I see that many of the colorful wooden beach huts (currently changing hands for as much as you would pay for a many-bedroomed house in some parts of the country) have been rebuilt since the recent storm. You can tell the old ones from the new ones quite easily. The steps up to the older huts and even their handrails are often buried beneath the sand that has mounded up at their feet. New huts have wisely been placed on foundation posts a meter or so higher than their neighbors.

In general, though, along this coast, the defenses held, evacuations

swung into action as planned, and nobody died. The morning after brought a mood, if not of smugness, then of satisfaction at a crisis well handled, tinged with some humility at the certain knowledge that it could easily have been worse. At Hopton-on-Sea, the new granite boulders passed their first severe test, although a stockpile of rocks waiting to be positioned in their planned locations had to be hurriedly requisitioned to save a section of cliff made exceptionally vulnerable by earlier storms. At Salthouse, a bulldozed shingle ridge normally protects the marshy polder behind from encroachment by the sea, and this wild expanse, in turn, acts as a sponge that provides an additional safeguard for the land behind. That night, the shingle was rudely breached, pushed far inland, with the sea pouring through the gaps it had made and across the marshes to flood the coast road and threaten properties. In the beach parking lot, the sign for the emergency telephone was to be found at ankle level on its post; the phone itself now lay under several feet of shingle. (Shingle laboriously piled up in calm weather by earth-moving equipment looks reassuring, but it can be deceptively protective—"purely placebo," in the words of Godfrey Sayers, a local adviser on the coastal environment. Unsurprisingly, perhaps, the natural profile established by the shingle over time is far more stable.)

In fact, a well-timed change of wind direction may have been the critical factor in preventing the sea from overtopping some flood barriers. Nevertheless, it seems unlikely that coastal communities will ever again be as unprepared as they were in 1953. A combination of better prediction and information sharing, and text messaging of flood warnings to civilians, will see to that. On the other hand, there is a growing tendency to regard extreme weather as an entertainment spectacle, as well as a sense that it is one's "right" as a consumer to be able to make a journey or do whatever else one pleases regardless of natural forces, and these hubristic attitudes may lead people to endanger themselves in new ways. Meanwhile, plans have been hastily drawn up for better sea defenses in urban areas, including Boston, King's Lynn, and Lowestoft; less densely settled coastal villages such as Hopton-on-Sea and Salthouse will increasingly have to fend for themselves.

In the safety of their laboratories, scientists from the UK Centre for Ecology & Hydrology, the Meteorological Office, and elsewhere later settled down to examine the question of whether the winter's storms—in addition to pluvial and fluvial flooding in much of the country, there were a number of coastal surges—could be firmly attributed to climate change or were merely extreme weather events within the bounds of what we should expect anyway. They concluded that it was impossible to know for sure, given the quality of present data and scientific modeling. They added that a major change in storm surge frequency was "unlikely over the coming decades." But the continued effect of rising sea levels during the next century and beyond (the sea-level commitment) means that, even if the climate were eventually to be stabilized, the frequency of extreme events can be expected to increase during that period—tenfold in many coastal locations, more than a hundredfold in some.

A FINAL EBBING

Soon after the 2013 storm, I return to the spot where I stood to make my measurements of the tide. My feet slowly impress themselves in the blue mud. The tidal flats are strewn with dead vegetation and boats that have broken away from their moorings. A channel marker buoy (starboard) lies capsized high on the marsh, like a green, unexploded bomb. The bridge where I fastened my tide gauge has been destroyed. It is the characteristic of the place that it is neither rockfast land nor permanently given to the slippery sea. The constant exchange between the two is what provides the *genius loci*. Today, the tide is held back by an easterly wind and rushes in as if it's late for an appointment—which in a way it is. But it makes it in the end, as it always does.

Is this the future? Is this east coast of mine a bellwether, an omen of environmental disaster and human doom, as poets and painters have so often portrayed it? In his collection entitled *An English Apocalypse*, the poet George Szirtes imagines five national apocalypses. One of the five is "Death by Deluge," in which a great tide strikes the East Anglian coast, sweeping away first the rump of Dunwich, then advancing, over the flats, up rivers, and across whole counties, to consume the high hills and ultimately the whole land. Is that the way it is? The latest scientific advice is, indeed, that we should remove sea defenses in some places in order to reduce the impact of coastal flooding.

So, should we read this yielding coast differently, with less antagonism, less as if it had some personal malevolence directed toward us? Shouldn't we learn to love its ever-changing shape and its constant capacity to surprise us? Doesn't it give as much as it takes?

Not far from where I am standing, on another stretch of fast-eroding Norfolk coast, a few months before that December storm surge, the routine action of the sea revealed a tantalizing glimpse of another lost world—not a prevision of our possible future this time, but a moment captured from our long-ago past. It exposed this vision on one high tide—and then promptly swept it forever from our gaze on another.

Early Pleistocene deposits at Happisburgh (pronounced "Haysbrough" or, more appropriately perhaps, "Haze-brrr") have long yielded evidence of flora and fauna that flourished in prehistoric times. Then, in 2005, the discovery of flint artifacts proved there was human occupation of northern Europe almost a million years ago, pushing back the known date by some 350,000 years. But as the erosion has continued, new and more astonishing knowledge has come to light.

Stormy weather in May 2013 exposed a new surface of laminated silt at the foot of the ever-retreating cliffs. An area about the size of a small bedroom was found to contain a series of softly rounded depressions unlike animal prints or the marks made by any natural abrasion. A team of archaeologists from the Pathways to Ancient Britain project was on site, about to embark on a routine geophysical survey, when one of their number, Martin Bates of the University of Wales Trinity Saint David, noticed the unusual traces and began to wonder whether they could be human footprints. Their position in the intertidal zone gave the scientists only three or four hours during each daytime low tide in which to study them—work that was further hampered by strong winds and lashing rain. The archaeologists just had time to take digital photographs of the scene that would furnish the precise measurements to make a thorough analysis possible. Already, the shapes were being erased little by little as they were covered by each tide. After a week, they were hard to discern at all. Then they were fully covered by the increasing spring tide and scrubbed out forever.

Fortunately, the photographs from this one session—taken using a camera that frequently had to be held under an umbrella while the tide was rising and the light was fading—proved sufficient for photogrammetric analysis. This detailed measurement procedure was able to reveal the order within the chaos of the depressions in the silt and to establish that they were indeed human. The team was able to date the prints using evidence from pollen trapped in the sediment and certain traces left by now extinct mammals. They were between 780,000 and 1,000,000 years old, making them the earliest hominin footprints to be found outside Africa.

The analysis revealed a mixed group of at least five adults and children, a family perhaps, walking along the tidal estuary of a large prehistoric river—in effect, the Thames, which then took a very different route to the sea. From the size of the prints, the scientists could estimate the stature of the members of the group, with the adults being comparable in height to average heights today, consistent with the early human species known as *Homo antecessor*, or "pioneer man."

Standing on the same spot, gazing out at the shallow brown sea, I find I am not so interested in their vital statistics. I wonder where those people were off to so long ago. Were they foraging for food in the intertidal zone? Or were they perhaps just enjoying a walk along the beach as we do today, with the children stopping off to marvel at the contents of the rock pools? Or do these footprints capture a moment in a more ambitious migration, forced upon these Norfolk predecessors of mine by their own environmental crisis? The pollen studies revealed vegetation typical of a cool climate, indicating that Happisburgh was close to the northern limit of human habitation in Europe at the time. The evidence suggests the group was heading south.

In the paper announcing their remarkable discovery, the archaeologists noted ruefully, "The rarity of such evidence is equalled only by its fragility at Happisburgh, where severe coastal erosion is both revealing and rapidly destroying sites that are of international significance." Well, that's the tide for you—creator and destroyer in one. Still, it's good to know that while the home owners on the cliff tops may be unhappy, the scientists on the beach below are eagerly looking forward to what further erosion will reveal.

GLOSSARY

amphidrome, amphidromic point   A location where there is no vertical tidal movement, around which the wave of the tide revolves. It may be an actual location at sea or on the coast, or a hypothetical point inland.

bore   The wave created when the rising tide surges rapidly up a river, especially one that is suitably funnel-shaped.

circatidal   Designating physiological activity in synchrony with the tides.

ebb, ebbing   The outgoing tide. Compare *flood, flooding*.

eddy   Flow in a direction against the main tide—for example, at the edges of a channel or in a bay.

flood, flooding   The incoming tide. Compare *ebb, ebbing*.

flow   The horizontal movement of the tide. Also, somewhat synonymous with *flood*.

harmonic   A mathematical term for quantities in reciprocal arithmetic progression (for example, 1, ½, ⅓, ¼, . . .).

intertidal Between the lines on the shore made by the high tide and the low tide.

neap A tide of relatively small range that occurs at half moon. Compare *spring*.

oscillation A regular to-and-fro movement, as of a pendulum, for example.

period The time taken to complete one oscillation. The inverse of frequency; thus, a twice-daily tide has a period of about twelve hours.

range The vertical distance covered by the water level between high and low tides. The range between average neap tides is less than between average spring tides. The greatest range is found with "astronomical tides," when certain moon, sun, and earth conditions coincide.

set The compass direction of tidal flow.

sine wave A mathematically pure oscillation in which amplitude is proportional to the sine of the phase angle (360 degrees being a complete cycle).

spring A tide of large range that occurs at new moon and full moon. Compare *neap*.

tidal atlas A set of charts of a body of water showing, usually by means of scaled arrows, the direction and rate of the tide at given intervals of time.

tide table A tabulation of the times of high (and often low) tides, often with relative heights. Usually published annually for a given set of ports.

ACKNOWLEDGMENTS

I was surprised when I first began to explore this topic to find no recent and accessible book devoted to the science of the tides. As I read more about the subject, and found myself assailed by unclear diagrams and intimidating equations, I began to see why this was so. This highly predictable daily and global phenomenon nonetheless has no single, simple cause; instead, its causes are numerous, intricate, and complicatedly linked. I realized that I would have to adopt a different strategy—to explain the scientific principles and the history of their discovery as best I could, but also to leaven this heavy bread with illustrations of the significance of the tides in the broader culture, and to introduce a much-needed physical sense of the sea's presence into a discussion that is often reduced to desiccated mathematical abstraction.

I enjoyed wide-ranging early conversations with a number of experts—conversations that may have seemed bewilderingly vague to them, but that certainly proved invaluable to me. For these I am grateful to Chris Jones and Tim Smith at the United Kingdom Hydrographic Office; to Kevin Horsburgh at the National Oceanography Centre; to John Mack at the Sainsbury Institute for Art; to John Aldridge, Julian Metcalfe, and Dave Hetherington at the Centre for Environment, Fisheries and Aquaculture Science; to Callum Roberts at the University of Hull; and to Rob Hall at the University of East Anglia.

I was guided on my travels in Nova Scotia by Carl Myers, Peter Smith, Phillip MacAulay, Chris Coolen, and Kent Smedbol at the Bedford Institute of Oceanography and the Canadian Hydrographic Service, as well as by Les Smith at the Annapolis Royal tidal-power station and by Mary McPhee at the Fundy Ocean Research Center for Energy; in the United States by Jerry Mitrovica and Harriet Lau at Harvard University, Richard Lindzen at MIT, Benjamin Carp at Tufts University, and Graham Giese, oceanographer emeritus at the Woods Hole Oceanographic Institution; in Norway by Roderick and Lindis Sloan, Lars and Therese Larsen, and Tor Tørresen at the Norwegian Mapping Authority Hydrographic Service, and Bjørn Gjevik at the University of Oslo; and in Venice by Giovanni Cecconi, Elena Zambardi, and Chiara Montan of Consorzio Venezia Nuova.

Other scientists from the many disciplines concerned with the sea generously supplied copies of their papers and patiently fielded my inexpert questions, among them Nick Ashton, Dario Camuffo, Marjorie Chan, Robert Dalrymple, Adriaan de Kraker, Martin Ekman, Rodney Forster, John Howe, Agustí Jansà, Charalambos Kyriacou, Bruce Levell, Susanna Seppala, Colin Shepherd, Paul Stancliffe, and Mikis Tsimplis.

I learned of other irresistible tidal stories from an equally broad range of sources, including Torquil Johnson-Ferguson, Robert Macfarlane, and Jules Pretty. The artists Susi Arnott, Crispin Hughes, Gayle Chong Kwan, and Andy Goldsworthy showed me how to look at the tide in different ways. Humphrey Berridge filled me in on the Battle of Maldon and kindly let me reproduce parts of his translation of the poem. Elizabeth James informed me about the rare tide clock on the tower of St. Margaret's Church in King's Lynn. Helen Johnston and Janita Drew led the Thames Discovery mudlarking expedition. George Wright and Lynne Bridge at the Environment Agency showed me around the Thames Barrier. Linda Barron, Anne Page, and Tim Miller led me to Shingle Street. Helen Francis kindly allowed me to join her charity walk across Morecambe Bay on behalf of the Royal National Lifeboat Institution, on which Cedric Robinson was a guide beyond compare. Mike Cowling told me about the Crown Estate's study of the changing British coastline through works of art. Orla Kennelly at the Norfolk Museums Service and Julia Orchard and Katy Barratt at the Royal Museums Greenwich told me more about some of the paintings in their collections. Grace Pitkethly, Geoff Davidson, Adrian Turpin, and Donna Brewster told me about the Wigtown Martyrs. Gaston Dorren told me more about the etymology of the word "maelstrom"; and Carl Kears, about the occurrence of tide terms in Old English manuscripts. Members of the Stiffkey Cockle Club (a sailing club, not a shell-fishing organization) regaled me with local tidal lore and introduced me to the eccentric science of the Holy Roman emperor Frederick II.

Many friends came up with other thoughts and leads, among them Jonathan and Laura Austin, Nick Bion, Will and Jane Carter, Ruth Garde, Jane Sears, Andrea Sella, Charlie and Helen Ward, and James and Sarah Wilding. As I was nearing the end of the manuscript, my American cousin David Redfield sent me some amusing cuttings that proved both a timely and a salutary reminder of the difficulty of my project.

I thank my agent, Antony Topping; my editors, Daniel Crewe and Matt Weiland; and copy editors Shan Morley Jones and Stephanie Hiebert.

My brother, John, helped me to recover memories of long-ago family sailing trips. Finally, thank you to Moira and Sam for everything.

BIBLIOGRAPHY

ACKROYD, PETER. *London: The Biography*. London: Chatto and Windus, 2000.

ADDISON, A. C. *The Romantic Story of the Mayflower Pilgrims*. London: Pitman, 1911.

ALDERSEY-WILLIAMS, HUGH. "Between Venice and the Deep Blue Sea." *New Scientist*, August 4, 1988, 45–52.

———. "Saving Venice." *Popular Science*, March 1988, 66–69.

ARNOTT, SUSI. *Estuary*. Walking Pictures, 2008. DVD.

ARRIAN. *The Campaigns of Alexander*. Translated by Aubrey de Sélincourt. Harmondsworth, UK: Penguin, 1971.

ASHTON, NICK, Simon G. Lewis, Isabelle De Groote, Sarah M. Duffy, Martin Bates, Richard Bates, Peter Hoare, et al. "Hominin Footprints from Early Pleistocene Deposits at Happisburgh, UK." *PLoS One*, February 7, 2014. doi:10.1371/journal.pone.0088329.

BAILEY, RODERICK. *Forgotten Voices of D-Day*. London: Ebury Press, 2009.

BALL, PHILIP. *H₂O: A Biography of Water*. London: Weidenfeld and Nicolson, 1999.

BARNES, R. S. K. *Coastal Lagoons*. Cambridge: Cambridge University Press, 1980.

———. *Introduction to Marine Ecology*. 2nd ed. Oxford: Blackwell, 1988.

BEEVOR, ANTHONY. *D-Day: The Battle for Normandy*. London: Penguin, 2009.

BELLOC, HILAIRE. *The Cruise of the Nona*. London: Century, 1983.

BOJER, JOHAN. *The Last of the Vikings*. Translated by Jessie Muir. London: Hodder & Stoughton, 1923.

BOUGAINVILLE, LOUIS-ANTOINE DE. *Voyage autour du monde*. Paris: Saillant & Nyon, 1771.

BROWNE, JANET. *Charles Darwin: A Biography*. Vol. 2, *The Power of Place*. London: Jonathan Cape, 2002.

BRYDONE, PATRICK. *A Tour through Sicily and Malta*. Perth, UK: R. Morison Junior, 1799.

BUNGE, J. H. O. *Tideless Thames in Future London*. London: Thames Barrage Association, 1944.

BYATT, A. S. *The Biographer's Tale*. London: Chatto and Windus, 2000.

CAMUFFO, DARIO. "Le niveau de la mer à Venise d'après l'oeuvre picturale de Véronèse, Canaletto et Bellotto." *Revue d'Histoire Moderne & Contemporaine* 57, no. 3 (2010): 93–110.

CAMUFFO, DARIO, and Giovanni Sturaro. "Sixty-cm Submersion of Venice Discovered Thanks to Canaletto's Paintings." *Climatic Change* 58 (2003): 333–43.

CARP, BENJAMIN L. *Defiance of the Patriots: The Boston Tea Party and the Making of America*. New Haven, CT: Yale University Press, 2010.

CARSON, RACHEL. *The Edge of the Sea*. Boston: Houghton Mifflin, 1955.

———. *The Sea Around Us*. New York: Oxford University Press, 1989.

———. *Under the Sea-Wind*. London: Penguin, 1996.

CARTWRIGHT, DAVID EDGAR. *Tides: A Scientific History*. Cambridge: Cambridge University Press, 1999.

CASHFORD, JULES. *The Moon: Myth and Image*. New York: Four Walls Eight Windows, 2003.

CHURCH, JOHN A., Philip L. Woodworth, Thorkild Aarup, and W. Stanley Wilson. *Understanding Sea Level Rise and Variability*. Chichester, UK: Wiley-Blackwell, 2010.

COOK, CAPTAIN JAMES. "On the Tides in the South Seas." *Philosophical Transactions of the Royal Society of London* 66 (1776): 447–49.

COOPER, J. A. G., and C. Lemckert. "Extreme Sea-level Rise and Adaptation Options for Coastal Resort Cities: A Qualitative Assessment from the Gold Coast, Australia." *Ocean & Coastal Management* 64 (2012): 1–14.

COOPER, W. S., I. H. Townend, and P. S. Balson, *A Synthesis of Current Knowledge on the Genesis of the Great Yarmouth and Norfolk Bank Systems*. London: Crown Estate, 2008.

Correspondence between the Society of Antiquaries and the Admiralty respecting the Tides in the Dover Channel, with Reference to the Landing of Caesar in Britain, B.C. 55; Together with Tables for the Turning of the Tide-Stream off Dover Made in the Year 1862. London: J. B. Nichols and Sons, 1863.

COULTON, G. G. *From St. Francis to Dante: Translations from the Chronicle of the Franciscan Salimbene*. Philadelphia: University of Pennsylvania Press, 1907.

CROSSLEY-HOLLAND, KEVIN. *The Penguin Book of Norse Myths*. London: Penguin, 1993.

Cruising Association Handbook. 8th ed. London: Cruising Association, 1996.

CUNLIFFE, BARRY. *Facing the Ocean: The Atlantic and Its Peoples, 8000 BC–1500 AD*. Oxford: Oxford University Press, 2001.

DAINTITH, JOHN, and Derek Gjertsen, eds. *A Dictionary of Scientists*. Oxford: Oxford University Press, 1999.

DARWIN, F., ed. *The Life and Letters of Charles Darwin*. London: John Murray, 1887.

DARWIN, G. H. *The Tides and Kindred Phenomena in the Solar System*. London: John Murray, 1898.

DAVEY, NORMAN. *Studies in Tidal Power*. London: Constable, 1923.

DAVIDSON, KEAY. *Carl Sagan: A Life*. New York: Wiley, 1999.

DAWSON, R. J., M. E. Dickson, R. J. Nicholls, J. W. Hall, M. J. A. Walkden, P. K. Stansby, M. Mokrech, et al. "Integrated Analysis of Risks of Coastal Flooding and Cliff Erosion under Scenarios of Long Term Change." *Climatic Change* 95 (2009): 249–88.

DEACON, MARGARET. *Scientists and the Sea*. London: Academic Press, 1971.

DEGREGORIO, SCOTT, ed. *The Cambridge Companion to Bede*. Cambridge: Cambridge University Press, 2010.

DE KRAKER, A. M. J. "Flooding in River Mouths: Human Caused or Natural Events? Five Centuries of Flooding Events in the SW Netherlands, 1500–2000." *Hydrology and Earth System Sciences* 19 (2015): 2673–84.

DENNY, M. W., and R. T. Paine. "Celestial Mechanics, Sea-Level Changes, and Intertidal Ecology." *Biological Bulletin* 194 (1998): 108–15.

DE ZOLT, S., P. Lionello, A. Nuhu, and A. Tomasin. "The Disastrous Storm of 4 November 1966 on Italy." *Natural Hazards and Earth Systems Science* 6 (2006): 861–79.

DRAKE, STILLMAN. "History of Science and Tide Theories." *Physis* 21 (1979): 61–69.

DUHEM, PIERRE. *Le système du monde: Histoire des doctrines cosmopologiques de Platon à Copernic*. Paris: Librairie Scientifique Hermann, 1954.

DYER, GEORGE C. *The Amphibians Came to Conquer: The Story of Admiral Richard Kelly Turner*. Washington, DC: US Department of the Navy, 1972.

EKMAN, MARTIN. *The Changing Level of the Baltic Sea during 300 Years: A Clue to Understanding the Earth*. Åland Islands, Finland: Summer Institute for Geophysics, 2009.

———. *An Investigation of Celsius' Pioneering Determination of the Fennoscandian Land Uplift Rate, and of His Mean Sea Level Mark*. Small Publications in Historical Geophysics 25. Åland Islands, Finland: Summer Institute for Historical Geophysics, 2013.

———. *The World's Longest Sea Level Series and a Winter Oscillation Index for Northern Europe 1774–2000*. Small Publications in Historical Geophysics 12. Åland Islands, Finland: Summer Institute for Historical Geophysics, 2003.

FALCONER, WILLIAM. *An Universal Dictionary of the Marine*. London: T. Cadell, 1771.

FEDERICO, MAURO, and Francesco Costanzo. "Predicting Marine Currents in the Strait of Messina." *Atti della Accademia Peloritana dei Pericolanti* 91, no. 1 (2013): A4.1–5.

FOX, CYRIL, and Bruce Dickins, eds. *Early Cultures of North-West Europe: H. M. Chadwick Memorial Studies*. Cambridge: Cambridge University Press, 1950.

GARRETT, CHRISTOPHER. "Tidal Resonance in the Bay of Fundy and Gulf of Maine." *Nature* 238 (1972): 441–43.

GILLIS, JOHN R. *The Human Shore*. Chicago: University of Chicago Press, 2010.

GILLISPIE, CHARLES C. *Pierre-Simon Laplace, 1749–1827: A Life in Exact Science*. Princeton, NJ: Princeton University Press, 1997.

GJEVIK, B., H. Moe, and A. Ommundsen. "Sources of the Maelstrom." *Nature* 388 (1997): 837–38.

———. "Strong Topographic Enhancement of Tidal Currents: Tales of the Maelstrom." Unpublished paper, University of Oslo, 1997.

GOSSE, EDMUND. *Northern Studies*. London: Walter Scott, 1890.

GREENBERG, JOHN L. *The Problem of the Earth's Shape from Newton to Clairaut*. Cambridge: Cambridge University Press, 1995.

GRIBBIN, JOHN. *Science: A History*. London: Penguin, 2002.

GRIEM, J. N., and K. L. M. Martin. "Wave Action: The Environmental Trigger for Hatching in the California Grunion *Leuresthes tenuis* (Teleostei: Atherinopsidae)." *Marine Biology* 137 (2000): 177–81.

GRIEVE, HILDA. *The Great Tide: The Story of the 1953 Flood Disaster in Essex*. Chelmsford, UK: County Council of Essex, 1959.

GRIFFIS, WILLIAM E. *Fairy Tales of Old Japan*. London: Harrap, 1908.

HALLIDAY, STEPHEN. *The Great Stink of London: Sir Joseph Bazalgette and the Cleansing of the Victorian Capital*. Thrupp, Stroud, Gloucestershire, UK: Sutton, 1999.

HAMILTON-PATERSON, JAMES. *Seven Tenths: The Sea and Its Thresholds*. London: Faber and Faber, 2007.

HANNING-LEE, F. C., and Admiralty Hydrographic Department. *West Coast of Scotland Pilot: Comprising the West Coast of Scotland from the Mull of Galloway to Cape Wrath including the Inner and Outer Hebrides*. 8th ed. London: HMSO, 1934.

HASKINS, C. H. "Science at the Court of Emperor Frederick II." *American Historical Review* 27 (1922): 670–94.

HEILBRON, J. L. *Galileo*. Oxford: Oxford University Press, 2010.

HOARE, PHILIP. *The Sea Inside*. London: Fourth Estate, 2013.

Hogwood, Christopher. *Handel: Water Music and Music for the Royal Fireworks*. Cambridge: Cambridge University Press, 2005.

Homer. *The Odyssey*. Translated by E. V. Rieu. London: Penguin, 2003.

Horner, R. W., and D. Clerk. "The Thames Barrier." *Proceedings of the Institution of Civil Engineers* 78 (1985): 15–25.

Hoskins, W. G. *The Making of the English Landscape*. Toller Fratrum, UK: Little Toller Books, 2013.

Howe, John, Riccardo Arosio, Dayton Dove, Roger Anderton, and Tom Bradwell. "The Seabed Geomorphology and Geological Structure of the Firth of Lorn, Western Scotland, UK." Paper presented at the EGU General Assembly, April 27 –May 2, 2014, Vienna, Austria.

Hughes, Paul. "Implicit Carolingian Tidal Data." *Early Science and Medicine* 8 (2003): 1–24.

Hugo, Victor. *Toilers of the Sea*. Translated by W. Moy. London: J. M. Dent, 1910.

Humboldt, Alexander von. *Cosmos: A Sketch of a Physical Description of the Universe*. Vol. 1. Translated by E. C. Otté and B. H. Paul. London: H. G. Bohn, 1849.

Humphreys, Colin. "Science and the Mysteries of Exodus." *Europhysics News*, May/June 2005: 93–96.

Hunter, John. "A Simple Technique for Estimating an Allowance for Uncertain Sea-Level Rise." *Climatic Change* 113 (2012): 239–52.

Huntingford, Chris, Terry Marsh, Adam A. Scaife, Elizabeth J. Kendon, Jamie Hannaford, Alison L. Kay, Mike Lockwood, et al. "Potential Influences in the United Kingdom's Floods of Winter 2013/14," *Nature Climate Change* 4 (2014): 769–77.

Intergovernmental Panel on Climate Change. *Climate Change 2013: The Physical Science Basis*. Geneva: IPCC Secretariat, 2013.

Inwood, Stephen. *The Man Who Knew Too Much: The Strange and Inventive Life of Robert Hooke 1635–1703*. London: Macmillan, 2002.

Jansa, A., S. Monserrat, and D. Gomis. "The Rissaga of 15 June 2006 in Ciutadella (Menorca), a Meteorological Tsunami." *Advances in Geosciences* 12 (2007): 1–4.

Jansen, Okka E., Geert M. Aarts, and Peter J. H. Reijnders. "Harbour Porpoises *Phocoena phocoena* in the Eastern Scheldt: A Resident Stock or Trapped by a Storm Surge Barrier?" *PLoS One*, March 6, 2013. doi:10.1371/journal.pone.0056932.

Kelvin, William Thomson, Baron, Joseph Larmor, and James Prescott Joule. *Mathematical and Physical Papers*. Cambridge: Cambridge University Press, 1911.

KEYNES, JOHN MAYNARD. *Essays in Biography*. New York: W. W. Norton, 1963.

KINGSHILL, SOPHIA, and Jennifer Beatrice Westwood. *The Fabled Coast*. London: Random House, 2012.

KOPPEL, TOM. *Ebb and Flow: Tides and Life on Our Once and Future Planet*. Toronto: Dundurn Press, 2007.

LAMB, HUBERT. *Historic Storms of the North Sea, British Isles and Northwest Europe*. Cambridge: Cambridge University Press, 1991.

LAVELLE, RYAN. *Alfred's Wars: Sources and Interpretations of Anglo-Saxon Warfare in the Viking Age*. Woodbridge, UK: Boydell Press, 2010.

LINNÉ, CARL VON. *Lachesis Lapponica, or A Tour in Lapland*. London: White and Cochrane, 1811.

LUBBOCK, BASIL. *The China Clippers*. Glasgow, UK: J. Brown and Son, 1914.

LYELL, CHARLES. "On the Proofs of a Gradual Rising of the Land in Certain Parts of Sweden." *Philosophical Transactions of the Royal Society of London* 125 (1835): 1–25.

LYNGE, BRIGIT KJOSS, Jarle Berntsen, and Bjørn Gjevik. "Numerical Studies of Dispersion Due to Tidal Flow through Moskstraumen, Northern Norway." *Ocean Dynamics* 60 (2010): 907–20.

MACGREGOR, DAVID R. *The Tea Clippers*. London: Conway Maritime Press, 1972.

MACK, JOHN. *The Sea: A Cultural History*. London: Reaktion, 2011.

MACMILLAN, D. H. *Tides*. London: CR Books, 1966.

MANDELBROT, B. B. "How Long Is the Coast of Britain? Statistical Self-Similarity and Fractional Dimension." *Science* 156 (1967): 636–38.

MARMER, H. A. *The Tide*. New York: D. Appleton, 1926.

MARSHALL, H. E. *Our Island Story*. London: T. C. and E. C. Jack, 1905.

MARTEN, MICHAEL. *Sea Change*. Heidelberg: Kehrer, 2012.

MARTIN, KAREN. "Introduction to Grunion Biology." 2006. http://grunion .pepperdine.edu/IntroductionToGrunionBiology.pdf.

MAWDSLEY, ROBERT J., Ivan D. Haigh, and Neil C. Wells. "Global Secular Changes in Different Tidal High Water, Low Water and Range Levels." *Earth's Future*, February 28, 2015. doi:10.1002/2014EF000282.

MAYHEW, HENRY. *London Labour and the London Poor*. New York: Dover, 1968.

McINNES, R., and H. Stubbings. *Art as a Tool in Support of the Understanding of Coastal Change in East Anglia*. London: Crown Estate, 2010.

McKIERNAN, PATRICK L. "Tarawa: The Tide That Failed." *US Naval Institute Proceedings* 88, no. 2 (February 1962): 38–49.

METCALFE, JULIAN D., Ewan Hunter, and Ainsley A. Buckley. "The Migratory Behaviour of North Sea Plaice: Currents, Clocks and Clues." *Marine and Freshwater Behaviour and Physiology* 39 (2006): 25–36.

MILLER, STANLEY L. "A Production of Amino Acids under Possible Primitive Earth Conditions." *Science* 117 (1953): 528–29.

MITROVICA, JERRY X., Natalya Gomez, and Peter U. Clark. "The Sea-Level Fingerprint of West Antarctic Collapse." *Science* 323 (2009): 753.

MOE, H., A. Ommundsen, and B. Gjevik, "A High Resolution Tidal Model for the Area around the Lofoten Islands, Northern Norway." *Continental Shelf Research* 22 (2002): 485–504.

MOORE, STUART A. *A History of the Foreshore and the Law Relating Thereto.* London: Stephens and Haynes, 1888.

MORTON, ALEXANDER S. *Galloway and the Covenanters.* Paisley, UK: Alexander Gardner, 1914.

NOF, DORON, and Nathan Paldor. "Are There Oceanographic Explanations for the Israelites' Crossing of the Red Sea?" *Bulletin of the American Meteorological Society* 73 (1992): 305–14.

O'REILLY, C. T., Ron Solvason, and Christian Solomon. "Resolving the World's Largest Tides." In *The Changing Bay of Fundy—Beyond 400 Years*, edited by J. A. Percy, A. J. Evans, P. G. Wells, and S. J. Rolston. Proceedings of the Sixth Bay of Fundy Workshop, Cornwallis, Nova Scotia (Environment Canada–Atlantic Region, Occasional Report, no. 23), 2005.

ORÓ, J., S. L. Miller, and A. Lazcano. "The Origin and Early Evolution of Life on Earth." *Annual Review of Earth and Planetary Sciences* 18 (1990): 317–56.

PAI, HSIAO-HUNG. "The Lessons of Morecambe Bay Have Not Been Learned." *Guardian*, February 3, 2014.

PALMIERI, PAOLO. "Re-examining Galileo's Theory of Tides." *Archive for History of Exact Sciences* 53 (1998): 223–375.

PARKER, BRUCE. *The Power of the Sea.* London: Palgrave Macmillan, 2010.

———, ed. *Tidal Hydrodynamics.* New York: Wiley, 1991.

———. "The Tide Predictions for D-Day." *Physics Today* 64, no. 9 (September 2011): 35–40.

PATRIDES, C. A., ed. *Sir Thomas Browne: The Major Works.* Harmondsworth, UK: Penguin, 1977.

PATTERSON, ARTHUR H. *Man and Nature on Tidal Waters.* London: Methuen, 1909.

PETTIGREW, JANE. *A Social History of Tea.* London: National Trust, 2001.

PHILBRICK, NATHANIEL. *Mayflower: A Story of Courage, Community, and War.* New York: Viking, 2006.

PICKERING, M. D., N. C. Wells, K. J. Horsburgh, and J. A. M. Green. "The Impact of Future Sea-Level Rise on the European Shelf Tides." *Continental Shelf Research* 35 (2012): 1–15.

PINCEBOURDE, SYLVAIN, Eric Sanford, and Brian Helmuth. "An Intertidal

Sea Star Adjusts Thermal Inertia to Avoid Extreme Body Temperatures." *American Naturalist* 174 (2009): 890–97.

PLINY THE ELDER. *Natural History: A Selection*. London: Penguin, 2004.

POE, EDGAR ALLAN. *The Science Fiction of Edgar Allan Poe*. Edited by Harold Beaver. Harmondsworth, UK: Penguin, 1976.

PONTOPPIDAN, ERICH, BISHOP OF BERGEN. *The Natural History of Norway*. London: A. Linde, 1755.

PRATER, A. J. "The Ecology of Morecambe Bay III. The Food and Feeding Habits of Knot (*Calidris canutus* L.) in Morecambe Bay." *Journal of Applied Ecology* 9 (1972): 179–94.

PRETOR-PINNEY, GAVIN. *The Wavewatchers' Companion*. London: Bloomsbury, 2010.

PRETTY, JULES. *This Luminous Coast*. Woodbridge, UK: Full Circle, 2011.

PUGH, DAVID. *Changing Sea Levels: Effects of Tides, Weather and Climate*. Cambridge: Cambridge University Press, 2004.

———. *Tides, Surges and Mean Sea-Level*. Chichester, UK: Wiley, 1987.

PUGH, DAVID, and Philip Woodworth. *Sea-Level Science: Understanding Tides, Surges, Tsunamis and Mean Sea-Level Changes*. Cambridge: Cambridge University Press, 2014.

PYE, MICHAEL. *The Edge of the World: How the North Sea Made Us Who We Are*. London: Penguin, 2014.

RABAN, JONATHAN. *Coasting*. London: Hodder and Stoughton, 1989.

———, ed. *The Oxford Book of the Sea*. Oxford: Oxford University Press, 1992.

———. *Passage to Juneau*. New York: Pantheon, 1999.

RAY, R. D., and P. L. Woodworth, eds. "Tidal Science in Honour of David E. Cartwright." Special issue, *Progress in Oceanography* 40, nos. 1–4 (1997).

REIDY, MICHAEL S. *Tides of History: Ocean Science and Her Majesty's Navy*. Chicago: University of Chicago Press, 2008.

REYNOLDS, A. T. "Capt. Wm. Hewett, R.N. and the Loss of H.M.S. 'Fairy.'" *Nautical Magazine* 10 (1841): 220–24.

RICKETTS, EDWARD F., and Jack Calvin. *Between Pacific Tides*. Stanford, CA: Stanford University Press, 1968.

Rivers and Tides: Andy Goldsworthy Working with Time. Directed by Thomas Riedelsheimer. London: Artificial Eye, 2009. DVD.

ROBERTS, MERVIN F. *The Tidemarsh Guide*. New York: E. P. Dutton, 1979.

———. *The Tidemarsh Guide to Fishes*. Old Saybrook, CT: Saybrook Press, 1985.

ROBINSON, CEDRIC. *Sandman*. Ilkley, UK: Great Northern Books, 2009.

———. *Time and Tide*. Ilkley, UK: Great Northern Books, 2013.

RUFUS, QUINTUS CURTIUS. *The History of Alexander*. Translated by John Yardley. Harmondsworth, UK: Penguin, 1984.

RUMBLE, ALEXANDER R., ed. *The Reign of Cnut: King of England, Denmark and Norway*. London: Leicester University Press, 1994.

SALIMBENE DE ADAM DA PARMA. *Cronica*. Edited by Giuseppe Scalia. Parma, Italy: Monte Università Parma, 2007.

SALLUST. *The Histories*. Translated by Patrick McGushin. Oxford: Oxford University Press, 1992.

SCHOFIELD, GUY. "The Third Battle of 1066." *History Today* 16 (1966): 688–93.

SEAR, D. A., S. R. Bacon, A. Murdock, G. Dongehan, P. Baggaley, C. Serra, and T. P. LeBas. "Cartographic, Geophysical and Diver Surveys of the Medieval Town Site at Dunwich, Suffolk, England." *International Journal of Nautical Archaeology* 40 (2011): 113–32.

SELLAR, W. C., and R. J. Yeatman. *1066 and All That*. London: Methuen, 1930.

SPRACKLAND, JEAN. *Strands: A Year of Discoveries on the Beach*. London: Jonathan Cape, 2012.

STEINBECK, JOHN. *The Log from the Sea of Cortez*. London: Penguin, 2001.

STEVENSON, ROBERT LOUIS, and Lloyd Osbourne. *The Ebb-Tide*. Edinburgh: Edinburgh University Press, 1995.

STOTHARD, PETER. *On the Spartacus Road*. London: Harper Press, 2010.

STRABO. *The Geography*. 8 vols. Translated by H. L. Jones. London: W. Heinemann, 1917–61.

TACITUS. *The Agricola and the Germania*. Translated by Harold B. Mattingley. London: Penguin, 2010.

TAFURI, MANFREDO. *Venice and the Renaissance*. Cambridge, MA: MIT Press, 1989.

TAYLOR, D. J. *Orwell*. London: Chatto and Windus, 2003.

TELFORD, MALCOLM. "A Hydrodynamic Interpretation of Sand Dollar Morphology." *Bulletin of Marine Science* 31 (1981): 605–22.

TESCH, FREDERICH-WILHELM. *The Eel*. 3rd ed. Edited by J. E. Thorpe. Translated by R. J. White. Oxford: Blackwell, 2003.

TESSMAR-RAIBLE, K., F. Raible, and E. Arboleda. "Another Place, Another Timer: Marine Species and the Rhythms of Life." *Bioessays* 33 (2011): 165–72.

THEROUX, PAUL. *The Happy Isles of Oceania*. London: Hamish Hamilton, 1992.

THOMPSON, SILVANUS PHILLIPS. *The Life of William Thomson, Baron Kelvin of Largs*. London: Macmillan, 1910.

THOMSON, DAVID. *The People of the Sea: Celtic Tales of the Seal-Folk*. Edinburgh: Canongate: 2001.

TOMASINO, M. "The Exploitation of Energy in the Straits of Messina." In *The Straits of Messina Ecosystem*, edited by L. Guglielmo, A. Manganaro, and E. De Domenico, 49–60. Messina, Italy: Università degli Studi di Messina, 1995.

TSIMPLIS, M. N. "Tides and Sea-Level Variability at the Strait of Euripus." *Estuarine, Coastal and Shelf Science* 44 (1997): 91–101.

UNITED KINGDOM HYDROGRAPHIC OFFICE. *West Coast of Scotland Pilot: West Coast of Scotland from Mull of Galloway to Cape Wrath including the Hebrides and Off-Lying Islands.* 17th ed. Taunton, UK: United Kingdom Hydrographic Office, 2011.

"Unusual Tides in the Falklands." MercoPress, January 10, 2005. http://en.mercopress.com/2005/01/10/unusual-tides-in-the-falklands.

VIBE, CHRISTIAN. *Arctic Animals in Relation to Climatic Fluctuations.* Copenhagen: Reitzel, 1967.

WADHAMS, STEPHEN. *Remembering Orwell.* Harmondsworth, UK: Penguin, 1984.

WALKER, BOYD W. "Periodicity of Spawning by the Grunion, Louresthes tenuis, an Atherine Fish." Scripps Institution of Oceanography Technical Report. PhD diss., University of California, 1949.

WATANABE, MASAKO. *Storytelling in Japanese Art.* New York: Metropolitan Museum of Art, 2011.

WESTFALL, RICHARD S. *Never at Rest: A Biography of Isaac Newton.* Cambridge: Cambridge University Press, 1980.

WHEWELL, W. "On the Tides of the Pacific, and on the Diurnal Inequality." *Philosophical Transactions of the Royal Society of London* 138 (1848): 1–29.

WHITE, MICHAEL. *Isaac Newton: The Last Sorcerer.* London: Fourth Estate, 1997.

WINCHESTER, SIMON. *Atlantic: The Biography of an Ocean.* London: Harper Press, 2010.

WODROW, ROBERT. *The History of the Sufferings of the Church of Scotland: From the Restauration to the Revolution.* Edinburgh: James Watson, 1721.

WOOTTON, DAVID. *Galileo: Watcher of the Skies.* New Haven, CT: Yale University Press, 2010.

ZALASIEWICZ, JAN. *The Planet in a Pebble: A Journey into Earth's Deep History.* Oxford: Oxford University Press, 2010.

ZANTKE, JULIANE, Tomoko Ishikawa-Fujiwara, Enrique Arboleda, Claudia Lohs, Katharina Schipany, Natalia Hallay, Andrew D. Straw, Takeshi Todo, and Kristin Tessmar-Raible. "Circadian and Circalunar Clock Interactions in a Marine Annelid," *Cell Reports* 5 (2013): 1–15.

ZHANG, LIN, Michael H. Hastings, Edward W. Green, Eran Tauber, Martin Sladek, Simon G. Webster, Charalambos P. Kyriacou, and David C. Wilcockson. "Dissocation of Circadian and Circatidal Timekeeping in the Marine Crustacean *Eurydice pulchra.*" *Current Biology* 23 (2013): 1–11.

INDEX